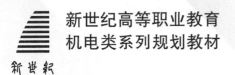

新世纪高等职业教育
机电类系列规划教材

机电英语（第三版）

附课文音频

主　编　李翠香
副主编　李　璟　宋婷婷　丁小月
　　　　龚晓霞　王春武

大连理工大学出版社

图书在版编目(CIP)数据

机电英语 / 李翠香主编. -- 3 版. -- 大连：大连理工大学出版社，2019.9(2024.7重印)
新世纪高职高专机电类课程规划教材
ISBN 978-7-5685-2012-6

Ⅰ. ①机… Ⅱ. ①李… Ⅲ. ①机电工程－英语－高等职业教育－教材 Ⅳ. ①TH

中国版本图书馆 CIP 数据核字(2019)第 103845 号

大连理工大学出版社出版

地址：大连市软件园路 80 号　邮政编码：116023
发行：0411-84708842　邮购：0411-84708943　传真：0411-84701466
E-mail:dutp@dutp.cn　URL:https://www.dutp.cn
大连永盛印业有限公司印刷　　大连理工大学出版社发行

幅面尺寸：185mm×260mm　　印张：11.5　　字数：264 千字
2009 年 3 月第 1 版　　　　　　　　2019 年 9 月第 3 版
2024 年 7 月第 5 次印刷

责任编辑：刘　芸　　　　　　　　责任校对：陈星源
封面设计：张　莹

ISBN 978-7-5685-2012-6　　　　　　　　定　价：38.80 元

本书如有印装质量问题，请与我社发行部联系更换。

前　言

　　《机电英语》(第三版)是新世纪高职高专教材编审委员会组编的机电类课程规划教材之一。

　　随着经济的国际化发展,世界上各种经济、文化、科技交往日趋频繁,而这一切均离不开语言这门工具,人们也越来越意识到英语在国际交流中的重要性,进而出现了大量以英语为专业的人才。但是,随着我国经济的腾飞,高技能型人才的缺乏越来越成为影响我国经济进一步快速、健康发展的瓶颈。

　　在加快高职高专教育改革和发展的过程中,教材的建设与改革起着至关重要的基础性作用,高质量的教材是培养高素质人才的保证。高职高专教材作为体现高职高专教育特色的知识载体和教学的基本工具,直接关系到高职高专教育能否为社会培养并输送符合要求的高技能型人才。

　　本教材以现代机电技术为背景,充分体现专业特色,顺应国家对培养机械相关专业技术人才的要求,满足企业对毕业生的技能需要,以服务教学、面向岗位、面向就业为方向,特邀请一批国内知名专家、学者、国家示范性高职院校的骨干教师和企业专家研读相关教材,讨论本教材上一版在使用过程中存在的不足和修订意见,力求为广大读者搭建一个高质量的学习平台。

　　在体系安排上,本教材既注重专业,又兼顾基础;既注重专业文章的阅读和理解,又兼顾专业词汇量的扩展;既注重专业性,又体现实用性。内容由浅入深,由易渐难。

　　本教材每个单元的内容分为翻译技巧和阅读训练两大部分。阅读训练部分共十个单元,每个单元包括三个部分:

　　第一部分为阅读理解,旨在培养学生阅读和理解机电专业英语的能力。该部分包含两篇文章,每篇文章的后面都有配套练习。

　　第二部分为模拟套写,旨在培养学生用英语模拟套写并翻译商业应用信件的能力,如广告、订单、保险单、简历等。

　　第三部分为会话,旨在培养学生运用相关英语进行涉外口语交际的能力。每个单元提供了四个情境对话实例,以供模仿。

全书最后附有专业英语的翻译技巧、翻译标准与方法及机电专业基础词汇表。

为方便教师教学和学生自学,本教材配有形式多样的立体化教学和学习资源,如有需要,请登录职教数字化服务平台进行下载。

本教材由平顶山工业职业技术学院李翠香任主编,平顶山工业职业技术学院李璟、宋婷婷、丁小月、龚晓霞及天津电气科学研究院有限公司王春武任副主编。具体编写分工如下:李翠香编写单元9、10;李璟编写单元7、8及附录1;宋婷婷编写单元5、6;丁小月编写单元3、4;龚晓霞编写单元1、2;王春武编写附录2、3。全书由李翠香负责统稿和定稿。

在编写本教材的过程中,我们得到了所在院校领导的高度重视与支持,同时还参考、引用和改编了国内外出版物中的相关资料以及网络资源,在此对这些资料的作者表示诚挚的谢意!请相关著作权人看到本教材后与出版社联系,出版社将按照相关法律的规定支付稿酬。

由于编者水平及时间所限,加上现代机电技术的不断发展,书中仍可能存在错误和疏漏,恳请各位读者谅解并不吝赐教。

编 者
2019年7月

所有意见和建议请发往:dutpgz@163.com
欢迎访问职教数字化服务平台:https://www.dutp.cn/sve/
联系电话:0411-84708979　84707424

目录

Unit 1　Machine Elements	1
Unit 2　Bearings and Shafts	13
Unit 3　Control Technology	26
Unit 4　Product Design	39
Unit 5　Modern Communication	52
Unit 6　Electric Technology	66
Unit 7　Inspection Technology	78
Unit 8　Development of Industrial Technology	90
Unit 9　NC Technology	102
Unit 10　Computer Control	116
参考文献	131
附　录	133

Unit 1
Machine Elements

Part 1 Reading

Passage A

Machine Elements

However simple, any machine is a combination of individual components generally referred to machine elements or parts. Thus, if a machine is completely dismantled, a collection of simple parts remains such as nuts, bolts, springs, gears, cams and shafts—the building block of all machinery.❶ A machine element is, therefore, a single unit designed to perform a specific function and capable of combining with other elements. Sometimes certain elements are associated in pairs,❷ such as nuts and bolts or keys and shafts. In other instances, a group of elements are combined to form a subassembly, such as bearings, couplings and clutches.

The most common example of a machine element is a gear, which, fundamentally, is a combination of the wheel and the lever to form a toothed wheel. The rotation of this gear on a hub or a shaft drives other gears that may rotate faster or slower, depending upon the number of teeth on the basic wheel.❸

Other fundamental machine elements have evolved from wheels and levers. A wheel must have a shaft on which it may rotate. The wheel is fastened to the shaft with couplings. The shaft must rest in bearings, and may be turned by a pulley with a belt or a chain connecting it to a pulley on a second shaft. The supporting structure may be assembled with bolts or rivets or by welding.❹ Proper application of these machine elements depends upon the knowledge of the force on the structure and the strength of the materials employed.

The individual reliability of machine elements becomes the basis for estimating the overall life expectancy of a complete machine.

Many machine elements are thoroughly standardized. Testing and practical experience have established the most suitable dimensions for common structural and mechanical parts. Through standardization, uniformity of practice and resulting economics are obtained. Not all machine parts in use are standardized, however. In the automotive industry, only fasteners, bearings, bushings, chains and belts are standardized. Crankshafts and connecting rods are not standardized.

New Words and Phrases

combination /ˌkɒmbɪˈneɪʃən/	n. 组合,结合
individual /ˌɪndɪˈvɪdʒuəl/	a. 单独的,各个的,个别的,特殊的
component /kəmˈpəʊnənt/	n. 元件,构件,部件
dismantle /dɪsˈmæntəl/	v. 分解(机器),拆开,拆卸
nut /nʌt/	n. 螺母
bolt /bəʊlt/	n. 螺栓;(门、窗等的)插销
spring /sprɪŋ/	n. 弹簧,板簧,簧片;弹力,弹性;v. 弹回,弹跳
gear /gɪə/	n. 齿轮,传动装置;v. 调整,(使)适合,换挡
cam /kæm/	n. 凸轮,偏心轮;样板,靠模,仿形板
shaft /ʃɑːft/	n. 柱身,连杆;传动轴
machinery /məˈʃiːnəri/	n. 机器
perform /pəˈfɔːm/	v. 执行,完成,做
associate /əˈsəʊsieɪt/	v. 联合,结合,参加,连带
key /kiː/	n. 键,电键,开关;楔;销;钥匙
subassembly /ˌsʌbəˈsembli/	n. 组合件,部件,机组
coupling /ˈkʌplɪŋ/	n. 联轴节,联轴器,联结器,联合器
clutch /klʌtʃ/	n. 离合器,夹紧装置
fundamentally /ˌfʌndəˈmentəli/	ad. 根本上
lever /ˈliːvə/	n. 杠杆,控制杆,操作杆
rotate /rəʊˈteɪt/	v. (使)旋转
hub /hʌb/	n. 轮毂
evolve /ɪˈvɒlv/	v. 进化,演变;开展,发展,展开
pulley /ˈpʊli/	n. 滑轮(组),滑车,带轮
assemble /əˈsembəl/	v. 安装,装配,组合,集合,集中
rivet /ˈrɪvɪt/	n. 铆钉;v. 铆接,铆

weld /weld/	v.&n.	焊接,熔焊
reliability /rɪˌlaɪə'bɪlɪti/	n.	可靠性,安全性,准确性
estimate /'estɪmeɪt/	v.	估计,估算,计算,测定,评价
expectancy /ɪk'spektənsi/	n.	期望,预期
thoroughly /'θʌrəl/	ad.	完全地,充分地,彻底地
standardize /'stændədaɪz/	v.	标准化,统一标准,标定,校准
establish /ɪ'stæblɪʃ/	v.	确定,制定,建立,创办,产生,使固定
dimension /daɪ'menʃən/	n.	尺寸,尺度,范围,方面
uniformity /ˌjuːnɪ'fɔːmɪti/	n.	均匀性,一致性
automotive /ˌɔːtə'məʊtɪv/	a.	自动的,汽车的
bushing /'bʊʃɪŋ/	n.	衬套,轴衬,轴瓦;[电](绝缘)套管
crank /kræŋk/	n.	曲柄
refer to		指的是,称为,涉及,关于
capable of		能够做的
combine with		与……结合

Notes

❶ Thus, *if* a machine is completely dismantled, a collection of simple parts remains such as nuts, bolts, springs, gears, cams and shafts—the building block of all machinery.

因此,如果把机床完全拆开,就可以得到像螺母、螺栓、弹簧、齿轮、凸轮及轴等简单零件——所有机器的组成元件。

"if"引导条件状语从句,表示"如果……"。

❷ Sometimes certain elements *are associated in pairs*…

有时某些特定的元件必须成对地工作……

be associated in pairs:成对地结合

例如:I am associated with him in business. 我与他两人合伙经商。

❸ The most common example of a machine element is a gear, *which*, fundamentally, is a combination of the wheel and the lever <u>to form a toothed wheel</u>. The rotation of this gear on a hub or a shaft drives other gears that may rotate faster or slower, depending upon the number of teeth on the basic wheel.

机械零件中最常见的是齿轮,它实际上是由轮子和杆组成的带有齿的轮子。在轮毂或轴上旋转的齿轮驱动其他齿轮做加速或减速运动,这取决于主动齿轮的齿数。

"which"引导非限制性定语从句,修饰"gear";不定式短语"to form a toothed wheel"作结果状语;分词短语"depending upon…"作原因状语。

❹ The supporting structure may *be assembled with* bolts or rivets or <u>by welding</u>.

支撑结构可由螺栓、铆钉或通过焊接固定在一起。

"be assembled with..."译为"由……安装";"by welding"作方式状语。

Exercises

Ⅰ. Answer the following questions according to the passage.

1. What is a machine element?

2. Which is the most common machine element?

3. What are other fundamental machine elements?

4. Which element is regarded as the basis for estimating the overall life expectancy of a complete machine?

5. Which machine elements are standardized in the automotive industry?

Ⅱ. Translate the following expressions into English or Chinese.

1. 机械零件
2. 简单零件
3. 自动化工业
4. 单一元件
5. 主动齿轮

6. the individual reliability of machine elements
7. the overall life expectancy of a complete machine
8. the building block of the machinery
9. the supporting structure
10. the strength of the materials

Ⅲ. Fill in the blanks with the proper expressions listed in the box. Change the form if necessary.

| standardize | reliable | machinery | component | uniform |
| expectancy | assemble | individual | gear | estimate |

1. Complex machines are made up of moving parts such as levers, _____, cams, cranks, springs, belts and wheels.

2. It can also be used to _____ the properties of the materials.

3. The _____ is required properties of the machine elements.

4. Agricultural Power _____ Operation is operating tractors and agricultural equipment.

5. A generator will read specifications of required variants, customize affected components and _____ them into a custom system.

6. You may use equipments, instruments or tools to identify bad engine _____.

7. Additional extensive testing was performed in both studies to evaluate product consistency and _____.

8. You should _____ costs and energy production using some basic assumptions.
9. It has a longer life _____ than concrete and steel equipment.
10. You can employ this method to analyze _____ electrical components of a complete machine for proper operation.

IV. Translate the following sentences into English.

1. 最常见的机械零件是齿轮,齿轮实际上是由轮子和杆组成的带齿的轮子。
2. 齿轮的硬度决定了它的耐磨性。
3. 制造业的工程师们集中精力研制标准化的零件。
4. 这些零件是在大批量、高规格和低成本的条件下生产的。
5. 可以说,零件表面是零件和设备的功能信息的载体。
6. 这些机械零件的正确使用与否,取决于是否懂得作用于结构上的力和所用材料的强度等相关方面的知识。

Passage B

Spur and Helical Gears

A gear having tooth elements that are straight and parallel to their axis is known as a spur gear. A spur gear can be used to connect parallel shafts only. Parallel shafts, however, can also be connected with gears of another type, and a spur gear can be mated with a gear of a different type.

To prevent jamming as a result of thermal expansion, to aid lubrication and to compensate for unavoidable inaccuracies in manufacture, all power-transmitting gears must have backlash. This means that on the pitch circles of a mating gear, the space width on the pinion must be slightly greater than the tooth thickness on the gear and vice versa. On instrument gears, using a gear split down its middle, one half being ratable relative to the others can eliminate backlash. A spring forces the split gear teeth to occupy the full width of the pinion space.

Helical gears have certain advantages. For example, when connecting parallel shafts, they have a higher load carrying capacity than spur gears with the same tooth numbers and cut with the same cutter. Because of the overlapping action of the teeth, they are smoother in action and can operate at higher pitch-line velocities than spur gears. The pitch-line velocity is the velocity in the pitch circle. Since the teeth are inclined to the axis of rotation, helical gears create an axial thrust. If used single, this thrust must be absorbed in the shaft bearings. The thrust problem can be overcome by cutting two sets of opposed helical teeth on the same blank. Depending on the method of manufacture, the gear may be of the continuous-tooth herringbone variety or a double-helical gear with space between the two halves to permit the cutting tool to run out. Double-helical gears are well suited for the efficient transmission of power at high speeds.

  Helical gears can also be used to connect non-parallel, non-intersecting shafts at any angle to one another. Ninety degrees is the commonest angle, at which such gears are used.

New Words and Phrases

spur /spɜː/	n. [建]凸壁;支撑物
helical /ˈhelɪkl/	a. 螺旋形的,螺旋线的
axis /ˈæksɪs/	n. 轴,轴线
mate /meɪt/	v. 使配对,使一致,结伴;n. 配偶,对手,助手
thermal /ˈθɜːməl/	a. 热的,热量的
expansion /ɪkˈspænʃn/	n. 扩充,开展,膨胀,扩张物,辽阔,浩瀚
lubrication /ˌluːbrɪˈkeɪʃn/	n. 润滑
compensate /ˈkɒmpənseɪt/	v. 偿还,补偿,付报酬
inaccuracy /ɪnˈækjʊrəsi/	n. 错误,误差
manufacture /ˌmænjʊˈfæktʃə/	v. 制造,加工;n. 制造,制造业,产品
backlash /ˈbæklæʃ/	n. 轮齿隙;反斜线(\);后座,后冲
pitch /pɪtʃ/	n. (齿轮)节距
pinion /ˈpɪnjən/	n. 小齿轮
split /splɪt/	v. 劈开,(使)裂开,分裂,分离;n. 裂开,裂口,裂痕
ratable /ˈreɪtəbl/	a. 可评价的,可估价的,按比例的
eliminate /ɪˈlɪmɪneɪt/	v. 排除,消除;除去
overlap /ˌəʊvəˈlæp/	v. (与某物)交叠,重叠,重合
velocity /vɪˈlɒsɪti/	n. 速度,速率,迅速,周转率
incline /ɪnˈklaɪn/	v. 使倾斜,赞同,喜爱
axial /ˈæksiəl/	a. 轴的,成轴的,轴向的
thrust /θrʌst/	n. [机]推力,侧向压力,插,猛推
herringbone /ˈherɪŋbəʊn/	n. 交叉缝式,人字形;a. 人字形的;v. (使)成箭尾形
transmission /trænzˈmɪʃn/	n. 传动,传递,变速器;[讯]发射,播送
intersect /ˌɪntəˈsekt/	v. 横切;横断;交叉,相交
spur gear	正齿轮
helical gear	斜齿轮
vice versa	反之亦然

Part 2 Simulated Writing

传真和电子邮件(Fax&E-mail)

与商业函件相比,传真和电子邮件不但快捷,而且比快件便宜,因此其应用范围越来越广。它们的格式与普通信函相似,但略有不同:传真信头中含有传真号码和电话号码,信内地址往往被收件人的传真号码和电话号码所代替;电子邮件则大多无信头内容,只有收件人的电子邮件地址及事由等。

一、传真

● Sample 1

VS(China)Electronics Co., Ltd.
GD Province, P. R. China 518527 Tel:86-755-27711288 Fax:86-755-27711289
Fax Cover Sheet

To: 至:	Sales Manager Philips Sound Systems (PSS) Corp.	From: 自:	Terry Madison Marketing Manager
Phone: 电话:	1-748-2152313	Phone: 电话:	86-755-27711288
Fax: 传真:	1-748-2152315	Fax: 传真:	86-755-27711289
Date of Issue: 发出日期:	Sep. 16,2019	Total Pages: 发出页数:	1 (Including this page)

Dear Sir or Madam,
 RE:Complaint about the Poor Quality of Woofers(低音喇叭,扩音器)
Please be informed that the woofers(PN(产品编号):4399 291 55236) you delivered on Sep. 15 were found with unacceptably high rate of broken cone paper(纸盆,喇叭中的一种物料)after sampling inspection by our QC inspectors, which caused considerable decrease in our daily production. We enclose the Inspection Report herewith, and have sent 5 defective samples to you. Please feed back your analysis report and improvement plan within one week. If this case recurs, we will have to reject later lots.

Yours faithfully,
Terry Madison
Marketing Manager

Sample 2

Herbert Import & Export.　　Telephone：(212)2215608　　Fax：(212) 2215706
Address：388 Station Street，New York，10018 U.S.A.
To：Johnson Company　　　　Date：May 23，2019
Attn：Eric Lee　　　　　　　　From：Simon Davis
Your Ref：2051/ef　　　　　　 Our Ref：5237/nl
CC：Kate Long
Page：1
Dear Sir，
We are a UK company; our shares are to be issued next month. Our company is to be granted a World Bank credit soon. Should you be interested in cooperation or buying our shares, please do not hesitate to contact us. We look forward to hearing from you.

Yours sincerely,
Simon Davis
Managing Director

传真格式要求

Ref(reference)—信函参考编号,一般编号往往包括有关人员姓名的首字母
Our Ref—我方编号,即发信人编号
Your Ref—贵方编号,即收信人编号
Attn—收件人
CC—抄送(其他收件人)
以上例文中的内容并非每一封传真都必须具备,而每一封传真都不能缺少的内容是：
To—收件人的姓名(后面可以加地址)
From(FM)—发件人的姓名(后面可以加地址)
Date—发传真的日期(月份一般用字母表示)
Fax—收件人的传真号码
Sub—事由
The salutation—称呼(和信函相同)
The body of the telex—电子邮件正文(和信函相同)
Regards—致意(和信函相同)
Signature—署名(和信函相同)

二、电子邮件

1. 电子邮件的格式

To—收件人的 e-mail 地址

Subject—主题

The salutation—称呼(和信函相同)

The body of the e-mail—电子邮件正文(和信函相同)

Regards—致意(和信函相同)

Signature—署名(和信函相同)

2. 英文电子邮件的基本写作技巧

英文电子邮件和英文信件差不多,主要由称呼、正文和落款三部分构成。写作正文时要注意以下两点:

(1)首先要明确写给谁。对待不同的收件人,语气会不同。写给朋友的可用一些俚语或缩略语,但是在比较正式的场合中,就不能用俚语或缩略语。

(2)写电子邮件要直接,并多用短句,使意思清楚。当然,对重点部分要做详细介绍。

Sample 3

To:Smith885@126.com

Subject:About myself

Dear Mr and Mrs Smith,

I'm a Chinese girl. My name is Yang Li. I'm very happy to know that I'll stay at your house for the English Summer Camp. I'd like to tell you something about myself so that you can know a bit about me before I arrive. I'm fifteen years old. I have a happy family with three people. I'm studying in No. 1 Middle School. We learn eight subjects. I'm interested in all of them. I like reading and playing the piano. I like English very much, but my English is not good enough. I think you can help me with my English. I hope to meet you soon.

Yours truly,
Yang li

Notes

❶ 在称呼处对方的名字不要拼错,头衔也不要错,头衔、学位任选其一。

❷ 人较多时,可用"Ladies and gentlemen"。

Practice

假如你叫方芳,在因特网上找到了一个叫 Joyce 的网友,现在请你根据下列要点给 Joyce 发 e-mail,介绍你自己的一些情况:

(1)方芳,中学生,家住在重庆;
(2)喜欢集邮和运动;
(3)对英语很感兴趣,会唱几首英文歌曲;
(4)班里同学也想找网友,希望得到帮助;
(5)想去英国看看。

Part 3　Speaking

Hometown and Education Background

(I＝interviewer 主试人　　A＝applicant 应试人)

● **Dialogue 1**

I：What's your permanent address?
A：My permanent address is Apt. 401, 126 Zhongshan Road, Nanjing.
I：Where is your birthplace?
A：My birthplace is Suzhou.
I：Are you a resident of Shanghai?
A：No.
I：Where is your domicile place?
A：My domicile place is Nanjing.
I：Give me your telephone number, please.
A：(My telephone number is) 8563361.

● **Dialogue 2**

I：What's your address?
A：My address is 356 Heping Road, Xuzhou.
I：Where are you working?
A：I'm working at Ping'an Hotel on Huaihai Road.
I：Where is your hometown?
A：My hometown is Suzhou.

I: Are you a local resident?
A: Yes, I am. I have been living in Suzhou since I graduated from university.

● Dialogue 3
I: Would you tell me something about your educational background?
A: Yes, sir. I graduated from high school in 2014, and then I entered Shanghai Polytechnics. I graduated in 2018, and I have a B. S. degree.
I: What department did you study in?
A: I was in the Department of Physics.
I: How were your scores at college?
A: They were all excellent.

● Dialogue 4
I: Which university did you graduate from?
A: Peking University. I had learned Economics there for four years.
I: What's your educational background?
A: I finished primary school in 2002, and entered middle school that September. I graduated from high school in July of 2008 and then I entered Peking University.
I: What are your major and minor subjects?
A: My major subject is Economics and my minor subject is English.
I: What course do you like best?
A: I am very interested in Business Management and I think it's very useful for my present job.
I: What do you think is the relationship between the subjects you have taken and the job you are seeking for?
A: I have taken courses on office administration, typing, reports and correspondence writing. Besides, I am also taking a Chinese type writing course. I think these are all closely related to a job of a junior secretary because it requires the ability to perform general office work and to assist the manager in handling all paper work.
I: How are you getting on with your studies?
A: I am doing well at school.
I: What subject are you least interested in?
A: I think it was Chinese History. Not because the subject was boring, but because of the large amount of materials that have to be memorized. It left me no room to appreciate the wisdom of great people in the past.
I: When and where did you receive your MBA degree?
A: I received my MBA degree from Peking University in 2018.

机电英语

New Words and Phrases

permanent /ˈpɜːmənənt/	a. 不变的, 永久的
resident /ˈrezɪdənt/	n. 居民
domicile /ˈdɒmɪsaɪl/	n. 户籍
local /ˈləʊkəl/	a. 当地的, 本地的
polytechnics /ˌpɒlɪˈtekniks/	n. 工业大学, 理工大学
major /ˈmeɪdʒə/	n. 主修课, 专业; a. 主要的, 较重要的
minor /ˈmaɪnə/	n. 副修科目; 未成年人; a. 较小的; 未成年的
administration /ædˌmɪnɪˈstreɪʃn/	n. 管理, 经营; 行政机关
correspondence /ˌkɒrɪˈspɒndəns/	n. 符合, 一致
B. S. degree＝Bachelor of Science degree	理学学士
primary school	小学
MBA＝Master of Business Administration	工商管理硕士

Useful Expressions

1. Where do you live? 你住哪里?
2. What's your address? 你的住址是哪?
3. I live at Apt. 401, 126 Zhongshan Road, Nanjing. 我住在南京市中山路126号401房间。
4. Could you tell me your telephone number? 能告诉我你的电话号码吗?
5. What is your birthplace? 你出生在哪里?
6. Where is your hometown? 你的籍贯是哪里?
7. Would you tell me what education background you have? 请告诉我你的学历好吗?
8. Which university did you graduate from? 你是哪个大学毕业的?
9. I'm a graduate of Shanghai Polytechnics. 我是上海理工大学毕业生。
10. What's your major in university? 你大学主修什么?
11. What are your major and minor subjects? 你的主修课和副修课都是什么?
12. I am very interested in Business Management and I think it's very useful for my present job. 我对企业管理非常感兴趣，而且我觉得它对我现在的工作很有帮助。
13. When and where did you receive your MBA degree? 你的 MBA 学位是什么时候、在哪里获得的?
14. I received my MBA degree from Peking University in 2019. 我于2019年在北京大学获得了 MBA 学位。
15. What do you think is the relationship between the subjects you have taken and the job you are seeking for? 你觉得你曾修读的科目和你申请的这份工作有什么关系?

Unit 2
Bearings and Shafts

Part 1 Reading

▶ **Passage A**

Types of Bearings

There are many types of bearings because of variations in the design of rings and separators and in the number of balls. ❶ They can be divided into classes according to their functions: those that support a radial, those that support a thrust load, or those that support a combination of thrust and radial loads.

1. Deep-groove Radial Ball Bearings

In deep-groove ball bearings, the races are approximately one-fourth as deep as the ball diameter. ❷ Although deep-groove ball bearings are designed to carry a radial load, they perform well under a combined radial and thrust load. For this reason, it is the most widely used type of ball bearings (Fig. 2-1).

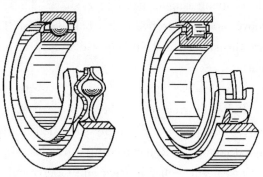

Fig. 2-1　Typical Ball Bearing

2. Filled-notch Ball Bearings

The most effective way of increasing bearing load capacity is to increase the ball complement. Filled-notch ball bearing is accomplished by cutting a notch in the races of a conventional bearing, thereby permitting the insertion of the balls. ❸ Once the balls have been inserted into the bearing through the notch, the notch is filled by an insert.

The insert should be kept on the unloaded side of the bearing, for it weakens the race. The increase ball complement makes possible a 20 to 40 percent increase in radial-load capacity over that of the normal deep-groove bearing, depending on the bearing size. The thrust capacity of the filled notch type is only 20 percent of the thrust capacity of the deep-groove bearing.

3. Angular Contact Bearings

A unidirectional thrust bearing, which is the most popular form of angular contact bearings, is common to us. This type of bearing is designed to support combined radial and unidirectional thrust loads. The amount of thrust load that this bearing can support is larger than the smaller contact angle.❹ A contact angle of zero corresponds to a radial bearing; a contact angle of 90 degrees corresponds to a thrust bearing.

4. Self-aligning Ball Bearings

Self-aligning ball bearings normally have two rows of balls that roll in a common spherical race in the outer ring.❺ Because of the design, the inner ring with the ball complement, can align itself freely around the axis of the shaft. When the shaft bends under load, the bearing will follow the deflection of the shaft without resistance, and self-alignment also contributes to smooth running by neutralizing the effect of the balls wobbling in the grooves. This bearing is therefore particularly useful in applications in which it is difficult to obtain the exact parallelism between the shaft and housing bores.

5. Thrust Ball Bearings

In the type of the simplest form of thrust ball bearings, a single row of balls set in a separator runs in two similar grooves formed in the stationary and revolving rings and the revolving ring is fixed to a shaft. These grooves are usually shallower than the groove in a deep-groove radial ball bearing. Thrust ball bearings are designed to carry pure thrust load and, if any radial load is present, separated radial bearing must be used.

6. Roller Bearings

Roller bearing serves the same purpose as ball bearings, but it can support much heavier load than a comparably sized ball bearing because it has a line contact instead of a point contact.❻ Roller bearing can be classified into three basic types: cylindrical roller bearing, needle roller bearing and tapered roller bearing.

New Words and Phrases

bearing /ˈbeərɪŋ/	n. 轴承；关系，方面，意义；(pl.)方向，方位
ring /rɪŋ/	n. 环，环状；铃声，打电话；v. 包围，按铃；响
separator /ˈsepəreɪtə/	n. 隔离物，离析器，脱脂器，分离者

Unit 2　Bearings and Shafts

radial /ˈreɪdɪəl/	a.	半径的,径向的;光线的,放射状的
groove /gruːv/	n.	(唱片等的)凹槽;惯例,最佳状态;v. 开槽于……
approximately /əˈprɒksɪmətli/	ad.	近似地,大约
notch /nɒtʃ/	n.	槽口,凹口;v. 刻凹痕,开槽;切口
complement /ˈkɒmplɪmənt/	n.	补足物;[数]余角;v. 补助,补足
accomplish /əˈkʌmplɪʃ/	v.	完成(某事物),做成功,实现
conventional /kənˈvenʃənəl/	a.	惯例的,常规的,习俗的,传统的
insertion /ɪnˈsɜːʃən/	n.	插入物,插入
unidirectional /ˌjuːnɪdɪˈrekʃənəl/	a.	单向的,单向性的
angular /ˈæŋɡjʊlə/	a.	有角的,有角度量的;(指性格)生硬的
correspond /ˌkɒrɪˈspɒnd/	v.	符合,协调,通信,相当,相应
aligning /əˈlaɪnɪŋ/	n.	列队,结盟
spherical /ˈsferɪkəl/	a.	球的,球形的,球状的,球面的
neutralize /ˈnjuːtrəlaɪz/	v.	使保持中立,使中立
wobble /ˈwɒbəl/	v.	摇晃,摆动,振动;n. 摆动,摇摆不定,颤动
parallel /ˈpærəlel/	a.	平行的;n. 平行线;v. (与某物)相当,相匹配
parallelism /ˈpærəlelɪzəm/	n.	平行,相同,相似,类似
revolve /rɪˈvɒlv/	v.	(使)旋转,考虑,循环出现
shallow /ˈʃæləʊ/	a.	浅的,浅薄的
cylindrical /sɪˈlɪndrɪkəl/	a.	圆柱的
taper /ˈteɪpə/	v.	使(某物)逐渐变窄,使(某物)逐渐变少

Notes

❶ There are many types of bearings **because of** variations in the design of rings and separators and in the number of balls.

由于内、外圈和保持架设计的多样化以及滚珠数目的不同,轴承有许多类型。

because of:介词短语,后接名词、动名词或动名词短语,作原因状语。

例如:Because of his wife being there, I said nothing about it.

❷ In deep-groove ball bearings, the races are **approximately one-fourth** as deep as the ball diameter.

在深沟径向球轴承中,滚道深度约为滚珠直径的四分之一。

approximately:相当于 about 或 nearly,即"大约"的意思。

one-fourth:四分之一,在句中用于比较级前,表示程度。分数中分子用基数词,分母用序数词,如分子大于1,则分母须加"s"。例如:two-fifths,译为"五分之二"。

as...as:像……一样,用于引出一个比较状语从句,两个"as"之间为形容词或副词。

❸ Filled-notch ball bearing is accomplished by cutting a notch in the races of a conventional bearing, thereby permitting the insertion of the balls.

在填充式球轴承中,通过在常规轴承的滚道上开一个槽,从那儿填充滚珠来增加滚珠的数量。

"cutting a notch in the races of a conventional bearing"是动名词短语,作介词"by"的宾语。

"thereby permitting the insertion of the balls"是现在分词短语,作伴随状语。

❹ The amount of thrust load that this bearing can support is larger than the smaller contact angle.

与小接触角轴承相比,这种轴承所能承受的推力载荷更大些。

"this bearing can support"是定语从句,修饰"load"。

❺ Self-aligning ball bearings normally have two rows of balls that roll in a common spherical race in the outer ring.

通常,自适应球轴承有两列滚珠,它们在外圈的公共球面滚道中滚动。

"that roll in a common spherical race in the outer ring"是定语从句,修饰"two rows of balls"。

❻ Roller bearing serves **the same** purpose **as** ball bearings, but it can support much heavier load than a comparably sized ball bearing because it has a line contact instead of a point contact.

滚子轴承与球轴承的功能一样,但其承受的载荷要远远大于相应尺寸的球轴承,因为它是线接触而不是点接触。

the same...as:引出比较状语从句,表示"与……一样",从句中省略了谓语动词"serve"或"do";"than a comparably sized ball bearing"是比较状语从句,从句中省略了谓语动词"supports"或"does"。

Exercises

Ⅰ. Answer the following questions according to the passage.

1. Do you know why there are so many types of bearings?

2. For what reasons are deep-groove radial ball bearings most widely used?

3. According to the passage, how many types can bearings be divided into? What are they?

4. Why can roller bearings support much heavier load than the same size ball bearings?

5. Depending on its size, after increasing ball complement, it is possible to make a 20 to 40 percent increase in radial-load capacity over that of the normal deep-groove bearing. Is it true?

_____

Ⅱ. Translate the following expressions into English or Chinese.

1. 轴承承载能力
2. 径向载荷
3. 滚子轴承
4. 轴向载荷
5. 公共球面滚道
6. deep-groove radial ball bearing
7. angular contact bearing
8. self-aligning ball bearing
9. thrust ball bearing
10. filled-notch ball bearing

Ⅲ. Fill in the blanks with the missing expressions according to the passage.

1. According to their function, bearings can be divided into radial loads, thrust loads and a _____ of thrust and radial loads.
2. The races are about one-fourth as deep as the ball diameter in _____ bearings.
3. To increase the ball complement can most effectively increase bearing _____.
4. Unidirectional thrust bearing is designed to _____ combined radial and unidirectional thrust loads.
5. When it is difficult to obtain the exact parallelism between the shaft and housing bores, _____ bearing is particularly commonly used.
6. The purpose of roller bearings is the same as that of ball bearings, but roller bearings can support much _____ than comparably sized ball bearings.
7. The most effective way of increasing bearing load capacity is to increase _____.
8. Self-aligning ball bearings normally have two rows of balls that roll in a common _____ in the outer ring.
9. In the type of the simplest form of thrust ball bearings, a single row of balls set in a separator runs in two similar grooves formed in the stationary and _____.
10. Roller bearing has a _____ instead of a point contact.

Ⅳ. Translate the following sentences into English.

1. 深沟球轴承是一种应用最广泛的球轴承。
2. 多数轴承需要进行定期维护，以避免过早毁损。
3. 滚子轴承承受的载荷要远远大于相应尺寸的球轴承。
4. 预期大约多达三分之二的材料将产生相同的效果。
5. 这一程序需要将处理器的能力提高25%。
6. 一般来说，轴承的表面温度与以往观测的一致。

Passage B

Shafts

Virtually all machines contain shafts. The most common shape for shafts is circular and the cross section can be either solid or hollow (hollow shafts can result in weight savings). Rectangular shafts are sometimes used, as in screwdriver blades, socket wrenches and control knob stems.

A shaft must have adequate torsional strength to transmit torque and not to be overstressed. It must also be torsionally stiff enough so that one mounted component does not deviate excessively from its original angular position relative to a second component mounted on the same shaft. Generally speaking, the angle of twist should not exceed one degree in a shaft length equal to 20 diameters.

A shaft is mounted on bearings and transmits power through such devices as gears, pulleys, cams and clutches. These devices introduce forces, which attempt to bend the shaft; hence, the shaft must be rigid enough to prevent overloading of the supporting bearings. In general, the bending deflection of a shaft should not exceed 0.01 in per ft of length between bearing supports.

In addition, the shaft must be able to sustain a combination of bending and tensional loads. Thus an equivalent load must be considered which takes into account both torsion and bending. Also, the allowable stress must contain a factor of safety, which includes fatigue, since torsional and bending stress reversals occur.

For diameters less than 3 in., the usual shaft material is cold-rolled steel containing about 0.4 percent carbon. Shafts are either cold-rolled or forged in sizes from 3 in. to 5 in.. For sizes above 5 in., shafts forged and machined to size. Plastic shafts are widely used for light load applications. One advantage of using plastic is safety in electrical applications, since plastic is a poor conductor of electricity.

Components such as gears and pulleys are mounted on shafts by means of key. The design of the key and the corresponding keyway in the shaft must be properly evaluated. For example, stress concentrations occur in shafts due to keyways, and the material removed to form the keyway further weakens the shaft.

If shafts run at critical speeds, severe vibrations can occur which can seriously damage a machine. It is important to know the magnitude of these critical speeds so that they can be avoided. As a general rule of thumb, the difference between the operating speed and the critical speed should be at least 20 percent.

Another important aspect of shaft design is the method of directly connecting one shaft to another. This is accomplished by devices such as rigid and flexible couplings.

New Words and Phrases

circular /ˈsɜːkjʊlə/	a. 圆形的,环形的
rectangular /rekˈtæŋgjʊlə/	a. [数]矩形的,长方形的,成直角的
overstress /ˌəʊvəˈstres/	v. 过压,过载
excessive /ɪkˈsesɪv/	a. 过度的,过分的,极度的
diameter /daɪˈæmɪtə/	n. 直径
rigid /ˈrɪdʒɪd/	a. 坚硬的,不弯曲的,严格的,坚强的
fatigue /fəˈtiːg/	n. 疲乏,疲倦,(金属材料的)疲劳
reversal /rɪˈvɜːsəl/	n. 反转,颠倒,(位子、功能等)转换
corresponding /ˌkɒrɪˈspɒndɪŋ/	a. 符合的,相应的,对应的,通信的
critical /ˈkrɪtɪkəl/	a. [物]临界的,危急的,关键性的
vibration /vaɪˈbreɪʃən/	n. 颤动,震动
magnitude /ˈmægnɪtjuːd/	n. 大小,量,数量,量值

Part 2 Simulated Writing

英文简历(English Résumé)

简历是对你自己简短的描述,说明你是哪儿的人、做过哪些事,也可以说简历是你自己给自己做的广告。英文简历能体现一个人的综合实力及英语水平。好的英文简历会给招聘方留下良好印象,从而为获得理想的工作奠定基础。

一、说明信

说明信是一种介绍性的信件,介绍你自己以及你应聘的目的,并概述你的技能和你能为公司做出的贡献。大多数公司除了需要简历之外,还要求附上一封说明信。

说明信主要包括三部分:开场白、自我介绍和结束语。

二、简历

一份正规的个人简历,要描绘出教育背景、专业目标、工作经历以及与专业相关的其他活动或参加的专业组织等。

英文简历与中文简历内容相似,主要包括以下几项基本内容:

个人资料:姓名、性别、出生日期、婚姻状况和联系方式等。

应聘职位：若将个人简历递交人才交流中心、劳务市场或劳务中介机构，则务必写出自己所希望获得的职位。例如：Assistant Manager，Secretary。

教育背景：从最高学历写起，一直向前推移。包括学校、专业和主要课程，所参加的各种专业知识和技能培训。

工作经历：按时间顺序列出参加工作至今所有的就业记录，包括单位名称、职务、就任及离任时间，应该突出所任每个职位的职责、工作性质等，此为求职简历的精髓部分。

其他：个人特长及爱好、奖励、其他技能、参加的专业团体、著述和证明人等。

三、简历的类型

好的简历应该写得既简短又内容丰富。目前英文简历主要有三种形式：按年月顺序编排的简历；按技能和能力编排的简历；混合形式的简历。下面分别以同一份简历为范例进行描述。

1. 按年月顺序编排的简历

按年月顺序是一种最常见的写简历的方式。它从最近的工作经历讲起，列举每一份工作的成绩，然后再列举教育背景，以时间为主线。实例如下：

● ANNA KING

- Address：15 Sample Rd Melbourne VIC 3000
- E-mail：a.king@jxue.com
- Home phone：(123)9999-1234
- Work phone：(123)9999-5678

● EMPLOYMENT HISTORY

- Marketing Manager（Melbourne 墨尔本）   2016-present time

The Wine and Food Emporium

Duties：

—Manage a staff of 18 people.

—Liaise(取得联系)with advertising agencies (above and below the line) and brief in all campaigns.

—Responsible for ＄15 million advertising budget and ＄80 million turnover.

- Marketing Research Manager 2013～2016

Di Pastio Pasta Products（Queensland）

Duties：

—Merchandise products in supermarkets—78 stores.

—Build in-store promotional displays (12 product lines).

● EDUCATION

- University of Queensland，Bachelor of Business(Marketing)，2012
- Seacliff TAFE，Associate Diploma in Marketing，2010

● COMPUTERS
- Platforms: Apple, IBM
- Software: Windows 7, Word, Excel, PowerPoint, PageMaker, Internet-trained
- Typing: 65 words per minute

● TRAINING
Train the Trainer Accreditation

● SPECIAL SKILLS
- Co-author of seven articles for "Marketing Management" magazine.
- Keynote speaker at the International MIA Annual Conference attended by 2 500 industry professionals.

2. 按技能和能力编排的简历
这种方式是根据工作经验,按照技能和能力来写简历。实例如下:

● ANNA KING
- Address: 15 Sample Rd Melbourne VIC 3000
- E-mail: a. king@jxue. com
- Work phone: (123)9999-5678

● OBJECTIVE
A position in senior marketing management with an internationally focused premium food, wine or produce company. Seeking to expand management and teambuilding skills and build solid brands throughout the world. Available to relocate locally or internationally.

● SUMMARY OF SKILLS
● MANAGEMENT

Managed a marketing team of 18 people, telemarketing teams of 75 people and a $15 million advertising budget. Responsible for the client/agency liaison between mainstream, below-the-line and data management agencies. Responsible for the overall profitability of five brands, four of which are market leaders in both share and volume.

● PRODUCT DEVELOPMENT

Responsible for the launch of two brands into the national market with each brand gaining a market share of 15 percent and 22 percent respectively within two years.

● SALES

Ground floor experience in sales and merchandising with international fast-moving packaged goods company. Territory Manager for North Western region covering 78 stores, 12 product lines and approximately 28 sales promotional events per year.

- STAFF TRAINING

Accredited "Train the Trainer" Instructor. Initiator of the Mentoring Programme at "Life Skills for Youth" organisation and now Board Member.

- PUBLIC SPEAKING

Keynote presenter at the Annual MIA Conference and regular guest speaker at CPI Award nights.

- COMPUTERS

Fluent in both Apple and IBM platforms; proficient in Excel, Word, PowerPoint, PageMaker, Typing(65 words per minute), Internet trained.

- PUBLISHED WORKS

Co-authored seven articles for the "Marketing Management" magazine.

- LANGUAGES

Fluent in French, both written and spoken.

WORK HISTORY
- 2016~Present Time: The Wine and Food Emporium, Melbourne
  Marketing Manager
- 2013~2016: Di Pastio Pasta Products, Queensland
  Marketing Research Manager

EDUCATION
- University of Queensland, Bachelor of Business(Marketing), 2012
- Seacliff TAFE, Associate Diploma in Marketing, 2010

REFERENCES
- Mr P Prentice, Di Pastio Products
  Ph: (07)999-7788
- Ms D Schwimmer, Faber Biscuits
  Ph: (03)999-4321

3. 混合形式的简历

把自己的优势如技能、经验和能力放在前面,然后按年月顺序列举工作经历和学历。实例如下:

ANNA KING
- Address: 15 Sample Rd Melbourne VIC 3000
- E-mail: a.king@jxue.com
- Mobile: 0412345678
- Work phone: (123)9999-5678

● OBJECTIVE
Senior Marketing Manager

● SUMMARY
Nine years in sales and marketing with a broad range of experience from ground-floor sales and merchandising to marketing management with an international producer and exporter of fine food and wine.

- **MANAGEMENT**

Managed a marketing team of 18 people, telemarketing teams of 75 people and was responsible for a $15 million advertising budget. Responsible for the client/agency liaison between mainstream, below-the-line and data management agencies. Responsible for the overall profitability of five brands, four of which are market leaders in both share and volume.

- **PRODUCT DEVELOPMENT**

Launched two brands onto the national market with each brand gaining a market share of 15 percent and 22 percent respectively within two years.

- **FINANCIAL**

Prepared quarterly and annual budget reports. Presented and reviewed the forecasts to senior management and represented the Australian management team at the International MIA Conference held in Chicago last December.

- **SALES**

Ground floor experience in sales and merchandising with an international fast-moving packaged goods company. Territory Manager for North Western region covering 78 stores, 12 product lines and approximately 28 sales promotional events per year.

● EMPLOYMENT HISTORY
Date: 2016～present time
Company: The Wine and Food Emporium
Title: Marketing Manager (Melbourne)
Duties: Responsible for a team of 18 people with an advertising budget of $15 million spread over 5 product lines. Chief liaison between clients and agencies and responsible for branding and product awareness. Increased turnover to $80 million in the last financial year, a 15 percent increase and was awarded the MIA's 2018 Best New Product. Generated over $200 000 of free trade journal publicity in the last year.

Date: 2013～2016
Company: Di Pastio Pasta Products(Sydney)
Title: Marketing Research Manager
Duties: Responsible for the management, co-ordination, recruitment and placement of 75 in-store demonstrators including an in-bound and out-bound telemarketing survey conducted in conjunction with the sampling demonstrations. Presented research findings

to the CEO level and was instrumental in the development of a new brand extension which resulted in a profit of ﹩2.1 million for the company.

● EDUCATION
- University of Queensland,Bachelor of Business(Marketing),2012
- Seacliff TAFE,Associate Diploma in Marketing,2010

● REFEREE
Available on request

Part 3 Speaking

Work

● Dialogue 1
A：May I help you,sir?

B：Yes. In yesterday's newspaper I saw that you're looking for a new product designer. Is that position still vacant?

A：Yes, we are looking for those with knowledge of developing modern mobile phones.

B：I'd like to apply for the job. Could I have an interview?

A：OK, I'll arrange it. Please fill in this application form first.

● Dialogue 2
A：Can you tell me how to write a résumé?

B：Oh, your name and address go at the top and also your phone number. Then you write what job you want. Make sure your résumé focuses on the kind of work you can do and want to do.

A：I see. What comes after that?

B：Work experience. You list your qualifications. Begin with your most recent experience first and work backwards.

A：Shall I include my titles or positions?

B：Of course. And list the years.

● Dialogue 3
A：Do you know the result of yesterday's interview?

B：Yes. I just heard from the manager of the personnel department. I didn't get the job.

A：Really? Don't worry too much about it. After all, it's your first time.

B：What can I do now?

A：You can try again, and I'll help you.
B：It's so kind of you. Thanks a lot.

● Dialogue 4

A：I was wondering whether you needed any part-timer.
B：What were you thinking of?
A：A hotel job of some sort.
B：Have you ever done anything similar?
A：No. Not so far.
B：There's nothing at present, but look back in a week.

New Words and Phrases

vacant /ˈveɪkənt/   a. 空的,空白的,未被占用的
qualification /ˌkwɒlɪfɪˈkeɪʃən/   n. 资格,合格证,合格证明

Useful Expressions

1. What do you do? 你从事什么工作?
2. What do you do for a living? 你以什么工作谋生?
3. What is your occupation? 你的职业是什么?
4. What type/kind of work do you do? 你从事何种工作?
5. Where do you work? 你在哪里工作?
6. I'm a salesman. / I'm in sales. 我是销售员。
7. How much salary do you expect? 你希望得到多少薪水?
8. What kind of salary are you hoping to get? 你希望得到何种薪水标准?
9. I hope the salary is 50 000 RMB per year. 我希望年薪为人民币 5 万元。
10. My salary requirement is in the $200 000～$350 000 range with appropriate benefits. 我想要的薪水为 200 000 至 350 000 美元,以及适当的福利待遇。
11. I'll leave that to you, sir. (薪水的事)那事由您决定,先生。
12. Money is important, but the responsibility that goes along with this job is what interests me most. 薪水固然重要,但这工作带来的责任更吸引我。
13. Can you speak Mandarin? 你能说普通话吗?
14. I can speak Mandarin fluently. 我能说流利的普通话。
15. Do you know any other languages? 你懂其他语言吗?

Unit 3
Control Technology

Part 1 Reading

Dimensional Control (I)

In the early days of engineering, the mating of parts was achieved by machining one part as nearly as possible to the required size, machining the mating part nearly to the size, and then completing its machining, continually offering the other part to it, until the desired relationship was obtained.❶ If it was inconvenient to offer one part to the other during machining, the final work was done at the bench by a fitter, who scraped the mating parts until the desired fit was obtained, the fitter therefore being a "fitter" in the literal sense. It was obvious that the two parts would have to remain together, and in the event of one having to be replaced, the fitting would have to be done all over again. In these days, we expected to be able to purchase a replacement for a broken part, and for it to function correctly without the need for scraping and other fitting operations.

When one part can be used "off the shelf" to replace another of the same dimension and material specification, the part is said to be interchangeable.❷ A system of interchangeability usually lowers the production costs, as there is no need for an expensive "fiddling" operation, and it also benefits the customers in the event of the need to replace worn parts. It also, however, demands that the dimensions of mating parts be specified, and that dimensional variations, due to machine and operator shortcomings, be taken into account. Some form of inspection must be introduced to ensure that the manufacture is controlled; this is particularly important, because dimensional errors may not be revealed until some time has elapsed, and often many miles away from the place where the machining was done.

Since it is accepted that it is virtually impossible to manufacture a part without error, or in the rare event of a part being without error,❸ to be able to proclaim it to be perfect (because the measuring instruments are subject to errors), it is necessary to indicate the maximum errors permitted. The draughtsman must indicate the largest and smallest sizes that can be permitted without the part functioning incorrectly. The extreme dimensions are called the limits of the size, and the difference between them is called the tolerance. The magnitude of the tolerance depends upon the type of operation involved, the skill of the machinist, the accuracy of the machine, and the size of the part. For a given grade of tolerance, the actual tolerance must be increased with the size. The tolerance should be as large as possible, to keep the cost to a minimum.❹

For the method of indicating, on a drawing, the permitted tolerance depends mainly upon the type of operation involved, but local preference must also be taken into account.❺ The following examples will illustrate some of the methods used.

- Unilateral limits. These are usually used when the distance between two faces, or the diameter of a hole or a shaft is specified. For example, when a diameter is being ground, the machinist would prefer to aim at the largest size permitted, so that, in the event of his reaching a diameter that is just a little larger than the maximum size permitted, he can take another cut, knowing that he can use up the whole of the tolerance before the job is rejected.❻ A draughtsman might dimension a nominal $75_{-0.012}^{0}$ mm diameter shaft as D75. Similarly, a nominal 75 mm hole might be dimensioned as $D75_{-0.012}^{0}$, the same reasoning applies for shafts.

- Bilateral limits. These are usually applied when, for example, the position of a hole is specified. The machine operator may position the hole nearer the datum or further from the datum than intended, and as the operator is in no position to change the situation when the hole has been started, he must aim between the limits of position, so that the maximum error can be made without causing the part to be rejected.❼ The center distance between two holes would therefore be specified as, for example, 100 ± 0.02 mm.

New Words and Phrases

dimensional /daɪˈmenʃənəl/	a.	尺度的,尺寸的
mating /ˈmeɪtɪŋ/	n.	配合
bench /bentʃ/	n.	钳工台
fitter /ˈfɪtə/	n.	装配工,钳工
scrape /skreɪp/	n. & v.	刮削

literal /ˈlɪtərəl/	a.	完全按照原文的,照字面本义的
fit /fɪt/	v. & n.	配合
specification /ˌspesɪfɪˈkeɪʃən/	n.	规格,规格说明;具体说明,详述
interchangeable /ˌɪntəˈtʃeɪndʒəbəl/	a.	可互换的,可交替的
system /ˈsɪstəm/	n.	系统;组合装置
fiddle /ˈfɪdəl/	v.	用小提琴演奏(曲调);胡乱摆弄
specify /ˈspesɪfaɪ/	v.	明确规定,评述,确切说明
elapse /ɪˈlæps/	v.	(时间)过去,消逝
tolerance /ˈtɒlərəns/	n.	容忍,宽容;公差,容许偏差
virtually /ˈvɜːtʃuəli/	ad.	事实上,实际上
proclaim /prəʊˈkleɪm/	v.	宣告,声明,显示
draughtsman /ˈdrɑːftsmən/	n.	绘图员
machinist /məˈʃiːnɪst/	n.	机械工,机械师
preference /ˈprefərəns/	n.	偏爱,优先选择
unilateral /ˌjuːnɪˈlætərəl/	a.	单方面的,单方面做出的;单边的
bilateral /ˌbaɪˈlætərəl/	a.	有两边的,双边的
datum /ˈdeɪtəm/	n.	论据,资料;基准
dimensional control		尺寸控制
nominal diameter		公称直径
unilateral limits		单边极限
bilateral limits		双边极限

Notes

❶ In the early days of engineering, the mating of parts was achieved by machining one part as nearly as possible to the required size, machining the mating part nearly to the size, and then completing its machining, continually offering the other part to it, until the desired relationship was obtained.

在早期的工程(问题)中,获得配合零件的方法是,首先尽可能把一个零件加工到所需尺寸,然后将与它配合的零件加工到接近所需尺寸,再不断将这两个零件试配,进一步加工直至获得所需配合的尺寸。

此句中,"by"后面跟了四个并列成分:"machining one part..."、"machining the mating part..."、"completing..."及"offering...",可翻译成"先……然后……再……进一步……"。

❷ When one part can be used "off the shelf" to replace another of the same dimension and material specification, the part is said to be interchangeable.

当一个零件从货架上拿来就可以替换同样材料规格的另一个零件时,就说明这种零件是可互换的。

❸ ... it is virtually impossible to manufacture a part without error, or in the rare event of a part being without error...

……事实上,零件不可能毫无误差地制造出来,或者说没有误差的零件是很少见的……

❹ For a given grade of tolerance, the actual tolerance must be increased with the size. The tolerance should be as large as possible, to keep the cost to a minimum.

对于给定级别的公差,实际公差须随着尺寸的增大而增大。为使加工成本最小,公差应尽可能取大一些。

❺ ... the permitted tolerance depends mainly upon the type of operation involved, but local preference must also be taken into account.

……许用公差主要取决于所用操作的类型,但也必须考虑本地的优先级。

❻ ... when a diameter is being *ground*, the machinist would *prefer to aim at* the largest size permitted, so that, in the event of his reaching a diameter that is just a little larger than the maximum size permitted, he can take another cut, knowing that he can use up the whole of the tolerance before the job is rejected.

……当直径要圆整时,机械工更愿意向最大允许尺寸圆整,这样,当他加工到所得直径尺寸略大于最大允许尺寸时,还可以在整个公差范围内再切一次,而且知道不会因此使产品报废。

这里"ground"是"grind"的过去式,译为"磨,削,圆整"。

"prefer to"译为"更喜欢,更愿意";"aim at"译为"朝着"。

❼ The machine operator may position the hole nearer the datum or further from the datum than intended, and *as* the operator is in no position to change the situation *when* the hole has been started, *he must aim* between the limits of position, *so that* the maximum error can be made without causing the part to be rejected.

机床操作人员可能将孔的位置定得较接近或远离所需数据,而且孔的加工一旦开始,操作人员便不可能改变孔的位置,他必须在尺寸位置限度内加工,以便在最大误差时不会使零件成为废品。

本句为并列复合句。"and"前面的分句是简单句,后面的分句是复合句。"as"引导原因状语从句,其中包含一个"when"引导的时间状语从句,"he must aim"是第二个分句的主干,"so that"引导目的状语从句。

Exercises

I. Answer the following questions according to the passage.

1. What is the way of parts mating in the early days of engineering?
   _____
2. What is the advantage of interchangeability system?
   _____
3. What is the disadvantage of interchangeability system?
   _____
4. What does the magnitude of the tolerance depend upon?
   _____
5. What should the tolerance be when we want to keep the cost to a minimum?
   _____

II. Translate the following expressions into English or Chinese.

1. 双边极限
2. 间隙配合
3. 过盈配合
4. 基轴制
5. 基孔制
6. tolerance and limits of size
7. a system of interchangeability
8. dimensional errors
9. maximum errors permitted
10. unilateral limits

III. Fill in the blanks with the proper expressions listed in the box. Change the form if necessary.

increase	clearance	specify	specification		hole-based
designate	indicate	transition	maximum errors permitted		shaft-based

1. The difference between the largest and the smallest sizes is called the _____.
2. For a given grade of tolerance, the actual tolerance must be _____ with the size.
3. When the limits of the size of both the hole and the shaft are such that the shaft is always smaller than the hole, the fit is said to be a _____ fit.
4. The fit between the clearance fit and the interference fit is a _____ fit.
5. When a standard shaft and a mating part whose size is larger or smaller than the hole are used to obtain the variation fit, it is said to be _____ system.
6. Either a shaft-based system or a _____ system may be used.
7. For any given basic size, a range of tolerances and deviations may be _____ with respect to the line of zero deviation, called the zero line.
8. The tolerance is a function of the basic size that is _____ by a number of symbols, called the grade—thus the tolerance grade.

9. The position of the tolerance with respect to the zero line, also a function of the basic size—is _____ by a letter symbol(or two letters), a capital letter for holes and a lowercase letter for shafts.

10. The _____ for a hole and shaft having a basic size of 45 mm might be 45 H8/g7.

Ⅳ. Translate the following sentences into English.

1. 显然,两个配件应该总在一起(工作),当其中任意一个需要替换时,所有的适配刮削工作又要从头重做。
2. 具有互换性的配件体系不必进行高成本的刮削操作,从而可降低生产的成本。当要换掉磨损的零件时,零件的互换性对客户而言也是很有益处的。
3. 必须采取某种形式的检测方法来确保对加工的控制,这一点特别重要。因为尺寸误差有时可能要过一段时间才会被发现,而此时却往往已远离加工的地方。
4. 这些尺寸的机制就称为极限尺寸,它们之间的差值称为公差。
5. 公差的大小依赖于所涉及的加工操作的类型、机械工的技能、机床的精确度以及零件的尺寸。
6. 单边极限通常用在两个面之间的距离、孔径或轴径被指定的情况下。

Passage B

Dimensional Control(Ⅱ)

Fits

Fits are concerned with the relationship between two parts. Consider a shaft and hole combination: if the shaft is larger than the hole, the condition is said to be of interference; and if smaller than the hole, the condition is said to be of clearance. The interference may be such that only shrinking can assemble the two parts, or it may be very slight, so that hand-operated press can assemble the parts. Similarly, the clearance can be slight, so that the shaft can rotate easily in the hole, or be large, so that there is ample clearance for bolts to pass through.

In order that the precise condition is ensured, the limits of the sizes of both the shaft and the hole must be stipulated.

● **Classes of Fit**

These are classified as follows:

Clearance fit. When the limits of the sizes of both the hole and the shaft are such that the shaft is always smaller than the hole, the fit is said to be a clearance fit.

Interference fit. When the limits of the sizes of both the hole and the shaft are such that the shaft is always larger than the hole, the fit is said to be an interference fit.

Transition fit. When the limits of the sizes of both the hole and the shaft are such that the condition may be clearance or interference, the fit is said to be a transition fit.

- **Hole-based Systems and Shaft-based Systems**

In order to obtain a range of degrees of clearance, and degrees of interference, it is necessary to use a wide variation of hole sizes and shaft sizes. For example, a manufacturing company could be making a number of parts, all of a nominal 25 mm diameter, but which are all slightly different in actual limits of the size, to suit the actual fit required of each pair of parts. This situation could mean that a large number of drills, reamers, gauges, etc., are required.

It is logical that, to reduce this number, a standard hole could be used for each nominal size, and the variation of fit be obtained by making the mating shaft smaller or larger than the hole, which is known as a hole-based system. Alternatively, a standard shaft could be used for each nominal size, and the variation of fit is obtained by making the mating hole larger or smaller, as required. This is known as a shaft-based system. A hole-based system is usually preferred, because it standardizes "fixed size" equipment such as reamer and plug gauges; but a shaft-based system is usually also provided, because sometimes it is more convenient to employ a common shaft to which a number of components are assembled, each with a different fit, and sometimes it is convenient to use bar stock without further machining.

Systems of Limits and Fits

It is convenient to establish a standardized system of limits and fits, not only to eliminate the need for the draughtsman to determine the limits each time an assembly is detailed, but also to standardize the tools and gauges required. A system of limits and fits should cater for a wide range of nominal sizes. To satisfy the various needs of industry, and should cater for a wide range of quality of work. The system should, if possible, be tabulated, to save the user the trouble of having to calculate the limits of the size to suit of the class of fit, the quality of the work, and the size of the part.

British Standard 4500: 2009, ISO Limits and Fits

This standard replaced BS 4500-5:1988, which was for both metric and inch sizes. Apart from being completely metric, BS 4500 is essentially a revision of BS 1988 to bring the British Standard into line with the latest recommendations of the International Organization for standardization(ISO). The system refers to holes and shafts, but these terms do not only be applied to cylindrical parts but can equally well be applied to the space contained or containing, two parallel faces or tangent places. The system is tabulated, and covers sizes up to 3 150 mm.

Unit 3　Control Technology

New Words and Phrases

shrink /ʃrɪŋk/	v.	(使)收缩,萎缩,退缩,畏缩
precise /prɪˈsaɪs/	a.	叙述清楚而准确的,精确的,独特的
stipulate /ˈstɪpjʊleɪt/	v.	讲明,规定(要求)
nominal /ˈnɒmɪnəl/	a.	名义上的,按计划进行的,列名的
reamer /ˈriːmə/	n.	[机]铰刀,铰床;果汁压榨机
gauge /geɪdʒ/	n.	标注尺寸,标注规格,规,量器,表
assembly /əˈsembli/	n.	集合,集会,装配车间,装配
tabulate /ˈtæbjʊleɪt/	vt.	把……制成表,列表显示,使成平面
metric /ˈmetrɪk/	a.	公制的,米制的
essentially /ɪˈsenʃəli/	ad.	必要地,本质地,基本上
tangent /ˈtændʒənt/	n.	切线,正切
clearance fit		间隙配合
interference fit		过盈配合
transition fit		过渡配合
hole-based system		基孔制
shaft-based system		基轴制
plug gauge		测量仪
bar stock		棒料
International Organization for Standardization(ISO)		国际标准化组织

Part 2　Simulated Writing

英文简历(English Résumé)

一、写作特点及简历范例

1. 突出重点

(1)充分体现你的优点,尤其是招聘方要求的知识、经验与技术。
(2)如实列出与招聘相关的业绩和成果。

2. 语言简洁

(1)通常要尽量省略主语"I",因为你的名字已经出现在个人资料栏目中。

(2)省略谓语动词"be"。例如:"I was born on January 6,1991."应改为"Born:January 6,1991."。

(3)应尽量用简单句。例如:"As I have been a typist for two years,I can type very quickly and accurately."应改为"Having been a typist for two years,I can type very quickly and accurately."。

(4)使用通用的缩略词。如:表示身高的"centimeter"可略为"cm",表示体重的"kilogram"可略为"kg"。

3. 篇幅适宜

略。

4. 避免差错

在书写简历时必须避免下列错误:

(1)语法错误和单词拼写错误;

(2)经历不完整;

(3)没有写明以前的工作业绩;

(4)版面设计混乱。

Résumé

Chinese Name:Guoqiang Zhang   English Name:Eddy Zhang

(外企习惯以英文名字作为同事间的称呼,如果你有英文名字,将会首先给你的面试官一份亲切感)

Sex:Male   Born:6/12/96   University:Peking University

Major:Marketing

Address:328#,Peking University

Telephone:1398***451

E-mail:***@163.com

- **Job Objective:**

A position offering challenge and responsibility in the realm of consumer affairs or marketing

- **Education:**

2014~2018:Peking University, College of Commerce

Graduating in July with a B.S. degree in Marketing

Fields of study include: economics, marketing, business law, statistics, calculus, psychology, sociology, social and managerial concepts in marketing, consumer behavior, sales force management, product policy, marketing research and forecast, marketing strategies

2008~2014:The No. 2 Middle School of Xi'an

(第二部分教育背景必须注意:求职者受教育的时间排列顺序与中文简历中的时间排列顺序正好相反,也就是说,是从求职者的最高教育层次写起)

- **Social Activities**：

2014～2018：Secretary of the Class League Branch

2008～2014：Class monitor

- **Summer Jobs**：

2016：Administrative Assistant in Sales Department of Xi'an Nokia Factory. Responsible for public relations, correspondence, expense reports, record keeping, inventory catalog

2017：Provisional employee of Sales Department of Xi'an Lijun Medical Instruments Equipment (Holdings) Company. Responsible for sorting orders, shipping arrangements, deliveries

- **Hobbies**：

Internet-surfing, tennis, travel

- **English Proficiency**：

College English Test Band Six

- **Computer Skills**：

Microsoft office, Adobe Photoshop, etc.

(大多数外企对英语及计算机水平都有一定的要求，个人的语言、计算机水平可单列说明)

二、注意事项

1. 写给谁看？

2. 他(她)喜欢看什么？

3. 言简意赅、段落清晰、易读易记。

4. 有内容、有深度、有风格。

5. 简历必须整洁，否则很容易被人忽略，涂涂抹抹、皱皱巴巴是很糟糕的，会给雇主留下坏印象。

6. 打印多份，甚至十几份、几十份以备用。

7. 要有针对性。如果对于不同的行业、不同的公司和不同的职位，你提交的都是同样的简历，那么这样的简历所欠缺的就是针对性。如果 A 公司要求你具备相关行业经验和良好的销售业绩，你在简历中清楚地陈述了有关的经历并且把它们放在比较突出的位置，这就是针对性；如果 B 公司要求你具备良好的英语口语能力，你在简历中描述了自己做过业余涉外商务翻译的经历，这就是针对性。

8. 客观性原则。简历上应该提供客观的可以证明你的资历、能力的事实、数据。比如："2001 年因销售业绩排名第一而获得公司嘉奖"。

Part 3 Speaking

Products

● **Dialogue 1**

A: Nice to meet you, Mr. Yang. I'm sorry to have kept you waiting.
B: Nice to meet you too, Miss Carter. Here's my card.
A: Thank you. Well, let's get down to business.
B: I have our latest catalogue with me. Look, this is the model you asked about. It is the best one available.
A: This looks very interesting.
B: We can offer you the same price for this one as last year's model.
A: Well, we'll have to do some comparing on our side.
B: I'm sure you'll be pleased with our products.

● **Dialogue 2**

A: It's kind of you to see me, Miss Stacy.
B: Feel free to call me Jane. I have heard a lot about you. I also thought you dealt with washing machines. Am I wrong?
A: No. We have been dealing with washing machines for years. We aren't completely satisfied with the quality, however.
B: We seldom get complaints on quality.
A: That's the reason I want to do business with you.
B: Is there any particular item you are interested in?
A: I think your model No. 1326 B is perfect for our needs.
B: We should have no problems supplying you.
A: There are certain areas I think we should discuss.
B: What sort of things would you like to discuss?
A: There are several questions.
B: Where should we start?
A: Let's begin with the price...

● **Dialogue 3**

A: I'd like to take this opportunity to discuss with you our yearly requirements of your woolen sweaters.
B: Yes, our woolen sweaters are in high demand in both domestic and international markets at this time of the year. May I know your requirements this time?

A: If you don't mind, I'd like to say a few words on the quality of your products.
B: Well, was there anything wrong with the quality of the woolen sweaters you bought last year?
A: There was no specific quality problem, but some customers complained that the sweaters were not as good as the ones they had bought before.
B: I'm shocked to hear that. Generally speaking, our products go through strict inspection procedures before warehousing and shipment.
A: I don't know how. Anyway the complaints do exist.
B: Oh. I feel terribly sorry. But I can assure you that our woolen sweaters are really up to the standard this time because they are all produced by a new advanced production line.
A: I'm glad to hear that. In this case, my requirements will still be the same as that of the last year.

Dialogue 4

A: It was very kind of you to give me a tour of the place. It gave me a good idea of your product range.
B: It's a pleasure to show our factory to our friends. What's your general impression, may I ask?
A: Very impressive, indeed, especially the speed of your NW Model.
B: That's our latest development. A product with high performance. We put it on the market just two months ago.
A: The machine gives you an edge over your competitors, I guess.
B: Certainly. No one can match us as far as the speed is concerned.
A: Could you give me some brochures for that machine? And the price if possible.
B: Right. Here are our sales catalogue and literature.
A: Thank you. I think we may be able to work together in the future.

New Words and Phrases

catalogue /ˈkætəlɒg/	n. 目录,目录册; v. 为……编目录
domestic /dəˈmestɪk/	a. 家庭的,国内的
woollen /ˈwʊlən/	a. 羊毛制的,毛线的,生产毛织品的; n. 毛织品,羊毛织物
procedure /prəˈsiːdʒə/	n. 过程,步骤,程序
warehouse /ˈweəhaʊs/	n. 货栈,仓库
exist /ɪgˈzɪst/	v. 存在

edge /edʒ/	n. 边缘
competitor /kəmˈpetɪtə/	n. 竞争者，比赛者，敌手
brochure /ˈbrəʊʃə/	n. 小册子

Notes

❶ get down to：开始做某事，密切关注某事
  例如：Let's get down to business.
       I like to get down to work by 9.
       It's time I got down to thinking about that essay.

❷ as far as ... is concerned：就……（方面）而言，在……方面

Useful Expressions

1. I'm sorry to have kept you waiting. 很抱歉让您久等了。
2. Well, let's get down to business. 那么，我们开始谈事情吧。
3. I have our latest catalogue with me. 我带来了我们最新的产品目录。
4. We can offer you the same price for this one as last year's model. 我们会按照去年的价格给您。
5. We'll have to do some comparing on our side. 我方还需要进行一些比较。
6. We seldom get complaints on quality. 我们极少听到关于产品质量的抱怨。
7. Is there any particular item you are interested in? 有特别感兴趣的产品吗？
8. There are certain areas I think we should discuss. 还有一些具体的细节有待商谈。
9. Let's begin with the price. 我们就先谈价格吧。
10. Our woolen sweaters are in high demand in both domestic and international markets at this time of the year. 每年这个时候，不论是国内市场还是国际市场，对我们的羊毛衫都有很大的需求量。
11. Generally speaking, our products go through strict inspection procedures before warehousing and shipment. 一般来说，我们的产品在入库和装运之前都经过了严格的检验程序。
12. What's your general impression, may I ask? 请问您的总体印象如何？
13. The machine gives you an edge over your competitors, I guess. 我认为这种机器使你比你的竞争对手多了一些优势。
14. No one can match us as far as the speed is concerned. 在速度方面，没有人能够与我们匹敌。

Unit 4
Product Design

Part 1 Reading

Injection Mold Design

The function of a mold includes two respects: imparting the desired shape to the elasticized polymer and cooling the injection molded part. ❶ It is basically made up of two sets of components: the cavities and cores and the base in which the cavities and cores are mounted. ❷ The mold, which contains one or more cavities, consists of two basic parts: a stationary mold half on the side where the plastic is injected; a moving mold half on the closing or ejector side of the machine. ❸ The separation between the two mold halves is called the parting line. In some cases the cavity is partly in the stationary and partly in the moving sections. The size and the weight of the molded parts limit the number of cavities in the mold and also determine the machinery capacity required. The mold components and their functions are as follows:

(1) Mold base—hold cavity (cavities) in fixed, correct position relative to machine nozzle.

(2) Guide pins—maintain proper alignment of two halves of mold.

(3) Sprue bushing (sprue)—provide means of entry into mold interior.

(4) Runners—convey molten plastic from sprue to cavities.

(5) Gates—control flow into cavities.

(6) Cavity (female) and force (male)—control size, shape and surface texture of mold article.

(7) Water channels—control temperature of mold surfaces to chill plastic to rigid state.

(8) Side (actuated by cams, gears or hydraulic cylinders)—form side holes, slots, undercuts and threaded sections.

(9) Vent—allow escape of trapped air and gas.

(10) Ejector mechanism (pins, blades, stripper plate)—eject rigid molded article from cavity or force.

(11) Ejector return pins—return ejector pins to retracted position as mold closes for next cycle.

The distance between the outer cavities and the primary sprue must not be so long that the molten plastic loses too much heat in the runner to fill the outer cavities properly. The cavities should be arranged around the primary sprue so that each receives its full and equal share of the total pressure available, through its own runner system (so-called balanced runner system).❶ This requires the shortest possible distance between cavities and primary sprue, equal runner and gate dimensions, and uniform cooling. Fig. 4-1 shows balanced and unbalanced cavity layouts.

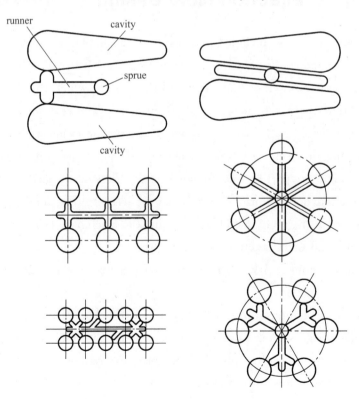

Fig. 4-1 Balanced and Unbalanced Cavity Layouts

New Words and Phrases

impart /ɪmˈpɑːt/	v.	将(某性质)给予或赋予某事物
elasticized /ɪˈlæstɪsaɪzd/	a.	用弹性线制成的
polymer /ˈpɒlɪmə/	n.	聚合物
cavity /ˈkævɪti/	n.	型腔,洞,中空
core /kɔː/	n.	型芯,核心,凹模,阴模
mount /maʊnt/	v.	安装,固定
ejector /ɪˈdʒektə/	n.	推出器,顶杆
separation /ˌsepəˈreɪʃn/	n.	分隔,分离
halve /hɑːv/	v.	二等分,平分
nozzle /ˈnɒzəl/	n.	喷嘴,浇口(把塑料从加热室注入模具处的终端)
alignment /əˈlaɪnmənt/	n.	校直,匹配,对中
interior /ɪnˈtɪərɪə/	n.	内部,里面
runner /ˈrʌnə/	n.	(塑料)流道
convey /kənˈveɪ/	v.	输送,传输
molten /ˈməʊltən/	a.	熔化的,熔融的
gate /geɪt/	n.	浇口(流道进入型腔处的终端)
force /fɔːs/	n.	凸模,阳模,力
texture /ˈtekstʃə/	n.	纹理,组织,结构
chill /tʃɪl/	v.	冷却
side /saɪd/	n.	侧抽芯机构,侧,侧柱
actuate /ˈæktʃueɪt/	v.	使(机器等)开动;发动
slot /slɒt/	n.	槽,直光道
undercut /ˌʌndəˈkʌt/	n.	切口,侧凹,凹割,根切
thread /θred/	n.	螺纹,丝线;v. 通过,穿过
vent /vent/	n.	通道,通风管,排气槽,孔口
blade /bleɪd/	n.	推板,刀片,螺旋桨,叶片
retract /rɪˈtrækt/	v.	缩回或拉回(某物);撤回或撤销(声名、指控等)
uniform /ˈjuːnɪfɔːm/	a.	形式或特征无变化的,一律的,均匀的;n. 制服
injection mold design		注射模具设计
impart sth. (to sb.)		将情况通知或告知(某人)
poly-		多,复,聚(前缀)
injection molded part		注射成型零件

stationary mold	定模
parting line	分型面,分型线
machinery capacity	机械(生产)能力,生产额
mold base	(动模、定模)底板,垫板
guide pin	导销,导柱
sprue bushing	主流道衬套(靠近注射机喷嘴),浇注套
water channel	冷却水通道
hydraulic cylinder	液压缸
stripper plate	分离装置,脱模机,脱模板
return pin	复位销,复位杆,回位销
primary sprue	主浇道,主流道
balanced runner system	对称(平衡)流道系统

Notes

❶ The function of a mold includes two respects: *imparting* the desired shape *to* the elasticized polymer and cooling the injection molded part.

模具的功能有两个方面:使可塑聚合物按希望的形状成型,并使其冷却定型。

"impart ... to ..."的含义是"将……传递给……,把……给予……"。

例如:His presence imparted an air of elegance to the ceremony.

他的出席给仪式增添了高雅的气氛。

❷ It *is* basically *made up of* two sets of components: the cavities and cores and the base in which the cavities and cores are mounted.

注射模具主要由两套部件组成:型腔和型芯以及安装型腔、型芯的模板。

be made up of:由……制成,由……拼成

"in which the cavities and cores are mounted"是定语从句,修饰"the base",介词"in"前置。

❸ The mold, which contains one or more cavities, consists of two basic parts: a stationary mold half on the side where the plastic is injected; a moving mold half on the closing or ejector side of the machine.

具有一个或多个型腔的模具由两个基本部分组成:注射塑料一侧的定模;闭合或顶出机构一侧的动模。

"which contains one or more cavities"是非限制性定语从句,修饰"mold"。"where the plastic is injected"是定语从句,修饰"side"。

❹ The cavities should be arranged around the primary sprue so that each receives its full and equal share of the total pressure available, through its own runner system (so-called balanced runner system).

型腔应安排在主流道的周围,这样各型腔能够通过各自的流道系统(称为平衡流道系统)分享充足的总压力。

"so that each receives its full and equal share of the total pressure available"为结果状语从句,"through its own runner system"是从句中的方式状语。

Exercises

Ⅰ. Answer the following questions according to the passage.

1. What is the function of a mold?

2. What is the mold mainly made up of?

3. What is the separation between stationary mold and moving mold?

4. What is the function of the vent as one of the mold components?

Ⅱ. Translate the following expressions into English or Chinese.

1. 侧面齿轮   6. locating ring
2. 主流道衬套   7. balanced runner system
3. 通风管   8. hydraulic cylinder
4. 主流道   9. side guide
5. 注射成型零件   10. injection mold design

Ⅲ. Fill in the blanks with the proper expressions listed in the box. Change the form if necessary.

cool	the parting line	stationary mold half	guide pin
distance	core	mold base	primary sprue
water channel	the size and the weight		

1. _____ can maintain proper alignment of two halves of a mold.

2. _____ holds cavity in fixed, correct, position relative to the machine nozzle.

3. The separation between the two mold halves is called _____.

4. The number of cavities in a mold is limited to _____ of the molded parts.

5. Two basic parts of a mold are _____ on the side where the plastic is injected and a moving mold half on the closing or ejector side of the machine.

6. Two sets of mold components are the cavities and _____ and the base in which the cavities and cores are mounted.

7. The shortest possible _____ is required between cavities and the primary sprue.

8. The function of a mold is twofold: imparting the desired shape to the elasticized polymer and _____ the injection molded part.

9. The cavities should be arranged around the _____ so that each receives its full and equal share of the total pressure available.

10. _____ control temperature of mold surfaces to chill plastic to rigid state.

IV. Translate the following sentences into English.

1. 塑料成型制品的尺寸和质量限制了一副模具中型腔的数目,并且确定了所需的注射机床的生产能力。
2. 有时型腔一部分在定模中,一部分在动模中。
3. 模板——保持型腔相对于注射机床的喷嘴具有固定、正确的位置。
4. 凹模(阴模)和凸模(阳模)——控制成型塑料制品的尺寸、形状及表面纹理。
5. 顶杆复位销——合模时使顶杆回到原始位置,为下一循环做准备。
6. 从外侧型腔到主流道的距离不能太远,否则熔融塑料在流道中损失的热量太多而不能有效地充满外侧型腔。

Passage B

Plastic Product Design

It is well known that plastics possess many valuable characteristics such as low weight, unlimited color ranges, esthetic values, low costs, and excellent mechanical, electrical and chemical properties, to name a few. The task of the designing work is to take the right combination of all these characteristics and embody them in the product to be molded.

First of all, the designer must consider what the articles are to accordingly determine the material, shape and process to be taken. Generally speaking, the following points should be noticed particularly in designing products:

(1) Keep wall thickness uniform. If the change in wall thickness is necessary, avoid

an abrupt change. Slope gradually from one thickness to the next.

(2) Include draft in all walls, ribs and bosses.

(3) Internal corners must be radiused at all times. External corners, if the design allows, should also be radiused.

(4) The rib thickness should be 60% to 70% of the adjoining wall thickness.

(5) Allow sufficient tolerance to compensate for variables in the material, the tool construction and the process.

(6) Avoid running threads to the end of the part. Allow clearance at both ends.

(7) Maintain close tolerance on mold seal off points of mold-in inserts.

(8) Provide inserts with proper anchorage to prevent pullout or rotation.

(9) Maintain adequate wall thickness over, around, and between inserts.

(10) Avoid locating holes too close to an edge.

(11) Holes to be trapped should be counter-sink.

(12) Use raised lettering in place of the depressed lettering where possible.

(13) Allow an additional draft for textured parts.

(14) Undercuts are costly to produce and should be avoided if possible.

New Words and Phrases

esthetic(al) /iːsˈθetɪk(əl)/	a.	[=aesthetic(al)]美学的;美的;审美的;艺术的
embody /ɪmˈbɒdi/	v.	体现,使具体化;包含,收录
abrupt /əˈbrʌpt/	a.	突然的,意外的
slope /sləʊp/	n.	斜线,斜面,山坡;v. 倾斜
draft /drɑːft/	n.	草案,草图;[机](拨模)斜度
rib /rɪb/	n.	肋条,排骨,棱纹,凸条
boss /bɒs/	n.	[机]铸锻件表面凸起的部分;轮毂
radius /ˈreɪdiəs/	n.	([复]radii 或 radiuses)辐射光线;范围
adjoining /əˈdʒɔɪnɪŋ/	a.	贴近的,毗邻的
clearance /ˈklɪərəns/	n.	[机]密封垫,焊接,封蜡,封印
insert /ɪnˈsɜːt/	v.	插入,嵌入;n. 插入物,插页
anchorage /ˈæŋkərɪdʒ/	n.	抛锚地点,停泊地点
pullout /ˈpʊlaʊt/	n.	拔,拉,撤离
countersink /ˈkaʊntəsɪŋk/	v.	钻(孔);穿(孔);使(钻头等)插入;n. 埋头钻
raised /reɪzd/	a.	凸起的;阳文的;有凸起花纹(或图案)的
lettering /ˈletərɪŋ/	n.	(写或刻印的)字
depressed /dɪˈprest/	a.	降低的;抑郁的,沮丧的

Part 2　Simulated Writing

英文招聘广告(English Employment Ads)

考虑当前的就业形势与特点,高职高专院校的学生在竞争激烈的就业市场中要想胜人一筹,就必须掌握阅读英文招聘广告这一英语应用技能。本文以下篇幅将讲解英文招聘广告的基本知识。

一、英文招聘广告的组成

英文招聘广告通常由以下内容组成:

1. 招聘单位的名称

招聘单位的名称一般单独列在第一行,排版时用粗体字,以标题的形式出现。例如:

Amway(China) Daily Necessities Company Limited
(安利(中国)日用品有限公司)

2. 招聘单位的标识

招聘单位的标识一般放在招聘广告标题的前面,也可以放在标题的后面。这是因为企业在招揽人才的同时,还要在读者的心目中树立起该企业的形象。

3. 招聘单位的简介

招聘单位的简介旨在对企业进行广告宣传,以便让求职者对企业有一个基本的了解。例如:

Krupp, one of the top ranking German industrial groups active in Mechanical Engineering, Plant Making, Automotive, Fabrication, Steel and Trading, is now widening its business in China.(克路普是德国名列前茅的工业集团公司之一,从事机械工程、工厂制造、汽车制造、装配、钢铁生产及贸易等业务,现正在扩大本公司在中国的业务。)

4. 招聘的职位

注明具体的空缺职位是任何招聘广告必不可少的内容。如果一则招聘广告只有一两项空缺,那么招聘的职位就往往用来做招聘广告的标题。例如:

VACANCY FOR OPERATORS OF NC MACHINE TOOLS(数控机床操作员空缺)

Job Opportunities:Typist & Office Clerk(就业机会:打字员和办公室职员)

5. 工作的职责

工作的职责是具体说明空缺职位应担任的主要职务,以便让求职者明确自己是否有能力胜任这些工作。例如:

Responsibilities(工作职责):

(1)Locating and appointing high caliber staff(寻找并聘用能力强的职员)

(2)Terms and conditions of employment(制订聘用条件)

(3)Career planning(制订工作计划)

(4)Training(培训员工)

6. 应聘的资格

应聘的资格即应聘的要求,也就是对应聘者在性别、年龄、学历、经验、特殊才能、个性、居住地等方面提出的具体要求。英语常用"qualifications""requirements"等来表达。例如:

Requirements(应聘要求):

(1)Experience working with foreign companies(有在国外公司工作的经验)

(2)Proficient in English and Mandarin(能熟练应用英语和普通话)

(3)Good at developing new business relationships(善于发展新的业务关系)

(4)Personal confidence and aggressiveness a must(必须有自信和进取心)

(5)Traveling within China a must(必须在国内出差)

(6)Mature, dynamic, honest(思想成熟,精明能干,为人诚实)

7. 提供的待遇

在国内报刊上刊登的英文招聘广告,对待遇问题一般都提得比较笼统。例如:

We offer an attractive salary package, fringe benefits and good opportunities for career development.(我们提供优厚的工资、福利和良好的职业发展机会。)

All positions offer highly competitive salaries, medical benefits and bonus, and of course, excellent training and career prospects.(所有职位都将获得颇具竞争力的工资、医疗福利和年度奖金,当然,还有良好的培训和职业前景。)

8. 应聘的方法

目前,国内的英文招聘广告绝大部分采用邮寄信件的方法招聘,要求应聘者邮寄个人简历、求职信、毕业证书和学位证书及身份证复印件、照片等资料,并写明邮寄地址及收件人。例如:

Successful applicants will be required to register through FESCO. Résumés together with a recent photograph and a contact telephone number should be sent to:

Office Manager

Rolls-Royce International Limited(China)

Room 14-15

International Building

19 Jianguomenwai Street

Beijing 100004

(应聘成功者必须在北京外企人力资源服务有限公司进行登记。个人简历(附近照一张及联系电话)须寄往:

100004 北京建国门外大街19号,国际大厦14-15房间,劳斯莱斯国际有限公司(中国)经理收)

二、英文招聘广告的语言特点

1. 创作新颖标题

为了吸引读者的注意,获得最佳的广告效果,广告撰稿人经常会别出心裁地遣词造句,创造出新颖标题。例如:

Hero Meets Hero(英雄识英雄)

It Is You Who Make Everything Possible(创造一切全靠你自己)

Are You Ready to Accept Challenges from a Transnational Enterprise?(你愿意接受跨国企业的挑战吗?)

2. 多用省略句

为了在有限的空间、时间、费用内传播足够的信息以达到最佳的广告效果,英文招聘广告大量使用省略句。例如:

Hard work and honesty a must.(工作勤奋和为人诚实是必备条件。)(主语之后省略了"is")

Read and write English.(会读、写英文。)(省略了情态动词"can")

Qualifications needed:(所需资格:)(省略了从句"which are")

3. 利用各种短语

一般用于说明工作职责和应聘资格(要求)两项内容。例如:

Good communication skills(良好的交际能力)

Ability to work in a team under pressure(能够在集体中承受压力进行工作)

4. 常用祈使句

一般用来说明应聘方法。为表示礼貌,使读者感到亲切,英文招聘广告的开头通常用"请"字。例如:

Please apply in your own handwriting with full English résumé and recent photo to:(请亲自誊写详细的英语简历并附上近照寄往:)

Please highlight the position you apply for at the bottom of the envelope.(请在信封下端注明你所应聘的职位。)

Please call 0510-5807123-3629 for interviewing.(请打电话0510-5807123-3629商定面试事宜。)

5. 采用缩略词

为了节省广告撰稿人的撰稿时间与读者的阅读时间以及广告的篇幅和费用,英文招聘广告中能缩略的词都尽可能缩略。例如:

Dept=Department(部门)   CV=Curriculum Vitae(个人简历)

JV=joint venture(合资企业)   G. M.=General Manager(总经理)

ad=advertisement(广告)   add=address(地址)

attn.=attention(请收信人注意)   ID=identification(身份)

Co. Ltd=corporation limited(有限公司)

三、英文招聘广告中常用的资格要求

1. 个人素质

A person with ability plus flexibility should apply.（需要有能力及适应力强的人。）

A stable personality and high sense of responsibility are desirable.（须个性稳重且具有高度责任感。）

Enthusiasm, organized working-habits are more important than experience.（有工作热情和有条不紊的办事习惯，经验不限。）

Being active, creative and innovative is a plus.（思想活跃，有创新精神尤佳。）

The main qualities required are preparedness to work hard, ability to learn, ambition and good health.（主要必备素质是吃苦耐劳精神、爱学习、事业心强和身体好。）

2. 语言能力

Ability to communicate in English is desirable.（能用英语进行交流。）

An excellent understanding of English would be mandatory.（具备出色的英文理解能力。）

Working command of spoken & written English is preferable.（有英文说、写应用能力者优先考虑。）

Able to speak Mandarin and the Cantonese.（会说普通话和粤语。）

3. 计算机知识

Computer operating skill is advantageous.（有计算机操作技能尤佳。）

Good at computer operation of Windows.（能熟练地在 Windows 下进行计算机操作。）

Familiar with CAD/CAM preferable.（熟悉 CAD/CAM 者优先。）

4. 工作经验

Working experience in foreign company is preferable.（有在外资公司工作经验者优先。）

At least two years' experience in operating NC machine tools.（至少有两年操作数控机床的经验。）

Familiarity with international trade issues will be an added advantage.（熟悉国际贸易问题者尤佳。）

5. 其他要求

Not more than 30 years.（年龄不超过 30 岁。）

Male/female.（男女均可。）

Part 3 Speaking

Work

◉ Dialogue 1

A: What are your greatest strengths and weak points?
B: I think I am very good at planning. But sometimes it sounds a little stubborn.
A: Oh, I see.
B: I'm proud of my social activities.
A: Can you work under pressure?
B: Yes, I find it stimulating.

◉ Dialogue 2

A: Excuse me.
B: Come this way, and please take a seat. My name is Bob Bush, the personnel manager of this company. May I have your name, please?
A: My name is Zhang Hanwei.
B: How did you hear about our company?
A: I read your recruiting advertisement in the newspaper.
B: What's your impression of our company?
A: I feel that your company has a lot of potential.

◉ Dialogue 3

A: May I help you?
B: Yes, I want to apply for the position as an office clerk.
A: I am Tony Brown, the manager of Human Resources Department. What's your name?
B: My name is Li Nan. How do you do, Mr. Brown?
A: I'm glad to meet you, Mr. Li. We have received your letter in answer to our advertisement. I would like to talk with you regarding your qualification for this interview.
B: I'm very happy that I'm qualified for this interview.

◉ Dialogue 4

A: Do you have any particular conditions that you would like our firm to take into consideration?
B: No, nothing in particular.
A: All right. If we decide to hire you, we would pay you 3 000 yuan a month at the start. You can have Saturday and Sunday off.
B: As regards salary, I leave it to you.
A: Well, we'll give you our decision in a couple of days.
B: Thank you!

New Words and Phrases

stubborn /ˈstʌbən/	a. 倔强的,顽固的,不屈不挠的,难处理的
stimulating /ˈstɪmjʊleɪtɪŋ/	a. 刺激的,鼓舞的,激励的,起促进作用的
be qualified for...	具有……资格
take into consideration	考虑,思考

Useful Expressions

1. What are your greatest strengths and weak points? 你最大的优点和缺点是什么?

2. I want to apply for the position as an office clerk. 我想申请办公室职员的职位。

3. We would pay you 3 000 yuan a month at the start. 最初我们将付给你每个月 3 000 元。

4. You can have Saturday and Sunday off. 你周六和周日可以休息。

5. As regards salary, I leave it to you. 关于薪酬,由您决定。

6. Can you talk about your education background? 能否谈一下你的教育背景?

7. Can you sell yourself in two minutes? Go for it. 你能在两分钟内进行自我推荐吗?大胆试试吧!

8. With my qualifications and experience, I feel I am hardworking, responsible and diligent in any project I undertake. Your organization could benefit from my analytical and interpersonal skill. 依我的资格和经验,我觉得我对所从事的每一个项目都很努力、负责、勤勉。我的分析能力和与人相处的技巧,对贵单位必有价值。

9. Give me a summary of your current job description. 对你目前的工作做个概括的说明。

10. I have been working as a computer programmer for five years. To be specific, I do system analysis, trouble shooting and provide software support. 我做了五年的计算机程序员。具体地说,我做系统分析、故障排除并提供软件支持。

11. I feel I can make some positive contributions to your company in the future. 我觉得未来我对贵公司能做些积极性的贡献。

12. What is your strongest trait? 你个性上最大的特点是什么?

13. Helpfulness and caring. 乐于助人和关心他人。

14. Adaptability and sense of humor. 适应能力和幽默感。

15. Cheerfulness and friendliness. 乐观和友爱。

Unit 5
Modern Communication

Part 1 Reading

▶ Passage A

5G

The transition to new fifth-generation cellular networks (Fig. 5-1), known as 5G, will affect how you use smartphones and many other devices. ❶ Let's talk about the essentials.

In 2019, a big technology shift will finally begin. It's an once-in-a-decade upgrade to our wireless systems that will start reaching mobile phone users in a matter of months. ❷

But this is not just about faster smartphones. The transition to new fifth-generation cellular networks — known as 5G for short — will also affect many other kinds of devices, including industrial robots, security cameras, drones and cars that send traffic data to one another. ❸ This new era will leap ahead of current wireless technology, known as 4G, by offering mobile Internet speeds that will let people download entire movies within seconds and most likely bring big changes to video games, sports and shopping.

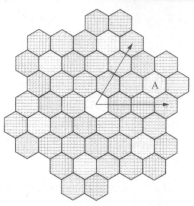

Fig. 5-1 Fifth-generation Cellular Networks

5G networks are regarded as a competitive edge. The faster networks could help spread the use of artificial intelligence and other cutting-edge technologies. ❹

● **What Exactly is 5G?**

Strictly speaking, 5G is a set of technical ground rules that define the workings of a cellular network, including the radio frequencies used and how various components like computer chips and antennas handle radio signals and exchange data. ❺

Since the first cellphones were demonstrated in the 1970s, engineers from multiple companies have convened to agree on new sets of specifications for cellular networks, which are designated a new technology generation every decade or so. To get the benefits of 5G, users will have to buy new phones, while carriers will need to install new transmission equipment to offer the faster service.

How Fast Will 5G Be?

The answer depends on where you live, which wireless services you use and when you decide to take the 5G plunge. ❻

Qualcomm, the wireless chip maker, said it had demonstrated peak 5G download speeds of 4.5 gigabits a second, but predicted initial median speeds of about 1.4 gigabits a second. That translates to roughly 20 times faster than the current 4G experience.

The 5G speeds will be particularly noticeable in higher-quality streaming videos. And downloading a typical movie at the median speeds cited by Qualcomm will take 17 seconds with 5G, compared with six minutes with 4G.

Rather than remembering to download a season of a favorite TV show before heading to the airport, for example, you could do it while waiting in line to board a plane, said the chip maker.

New Words and Phrases

transition /træn'zɪʃən/	n.	过渡,转变,变迁;[语]转换;[乐]变调
cellular /'seljʊlə/	a.	细胞的;蜂窝状的;由细胞组成的;多孔的
smartphone /'smɑːtʃəʊn/	n.	智能手机
essential /ɪ'senʃəl/	a.	必要的;本质的;基本的;精华的;n. 必需品;基本要素
decade /'dekeɪd/	n.	十年,十年间;十个一组;十年期
upgrade /ʌp'greɪd/	v.	提升;使(机器、计算机系统等)升级;n. 向上的斜坡
drone /drəʊn/	v.	嗡嗡叫;絮絮叨叨地说;n. 雄蜂;无人驾驶飞机;嗡嗡声
era /'ɪərə/	n.	纪元,年代;历史时期,时代;重大事件
leap /liːp/	v.	跳;冲动的行动;跳过,跃过;使跳跃;n. 跳跃,飞跃
download /'daʊnləʊd/	v.	下载
competitive /kəm'petɪtɪv/	a.	竞争的,比赛的

机电英语

artificial /ˌɑːtɪˈfɪʃəl/	a. 人工的；人造的；n. 人造肥料；假花
define /dɪˈfaɪn/	v. 规定；使明确；精确地解释；下定义
antenna /ænˈtenə/	n. 天线；[生]触角，触须；感觉，直觉
multiple /ˈmʌltɪpəl/	a. 多重的；多功能的；n. 倍数；[电工学]并联
convene /kənˈviːn/	v. 召集；聚集；传唤；集合
specification /ˌspesɪfɪˈkeɪʃən/	n. 规格；说明书；详述
designate /ˈdezɪɡneɪt/	v. 指派；指出；把……定名为；a. 指定而尚未上任的
carrier /ˈkærɪə/	n. 搬运人；运输公司；搬运器；[医学]带菌者
install /ɪnˈstɔːl/	v. 安装；安顿，安置；任命
plunge /plʌndʒ/	v. 用力插入；使陷入；跳入；俯冲；n. 投入；游泳
gigabit /ˈɡɪɡəbɪt/	n. [计]千兆比特
initial /ɪˈnɪʃəl/	a. 最初的；开始的；首字母的；n. 首字母；[语音学]声母
median /ˈmiːdiən/	a. 中间的；中央的；n. 中位数；中线；[数]中值
roughly /ˈrʌfli/	ad. 粗略地；大体上；粗暴地
noticeable /ˈnəʊtɪsəbəl/	a. 显而易见的；令人瞩目的；显著的；可以察觉的
cite /saɪt/	v. 引用，引证；表扬；[军事]传（或通）令嘉奖
transition to	向……过渡，向……转变
known as	被称为，公认为

Notes

❶ The transition to new fifth-generation *cellular networks* (Fig. 5-1), known as 5G, will affect how you use smartphones and many other devices.

向新的第五代蜂窝网络（图5-1，被称为5G）的过渡将影响你如何使用智能手机和许多其他设备。

"5G"全称是"fifth-generation cellular networks"，即第五代蜂窝网络。

蜂窝网络（cellular networks）又称移动网络（mobile networks），是一种移动通信硬件架构。蜂窝网络被广泛采用源于一个数学猜想，正六边形被认为是使用最少节点可以覆盖最大面积的图形，出于节约设备构建成本的考虑，正六边形是最好的选择。这样形成的网络覆盖在一起，形状非常像蜂窝，因此称其为蜂窝网络。

"cellular"还译为"细胞的；由细胞组成的"。例如：cellular structure（细胞结构）。

❷ It's an *once-in-a-decade* upgrade to our wireless systems that will start reaching mobile phone users *in a matter of* months.

这是我们无线系统十年一次的升级，大约几个月后开始覆盖手机用户。

once-in-a-decade：十年一次的。例如：once-in-a-decade leadership transition（十年一次的领导层换届）。

in a matter of：左右，大约，大约只不过在……之内。

❸ The transition to new fifth-generation cellular networks — known as 5G for short — will also affect many other kinds of devices, including industrial robots, security cameras, drones and cars that send traffic data to one another.

向新的第五代蜂窝网络（简称5G）的过渡，还将影响许多其他类型的设备，例如工业机器人、监控摄像机、无人机以及相互发送交通数据的汽车。

❹ The faster networks could help spread the use of artificial intelligence and other *cutting-edge* technologies.

更快的网络有助于推广人工智能及其他尖端技术。

edge：(微弱的)优势。例如：The company needs to improve its competitive edge. /The next version of the software will have the edge over its competitors.

cutting-edge：尖端的，先进的。

❺ *Strictly speaking*, 5G is a set of technical *ground rules* that define the workings of a cellular network, including the radio frequencies used and how various components like computer chips and antennas handle radio signals and exchange data.

严格地说，5G是一套定义蜂窝网络工作原理的基本技术规则，包括使用的无线电频率以及计算机芯片和天线等各种组件处理无线电信号和交换数据的方式。

strictly speaking：严格来说，严格地说。类似词组还有 frankly/generally/honestly speaking。

ground rules：基本规则，基本原则。lay down/establish ground rules for frequency 作科技术语时表示"(无线电波、声波等的)频率"。

❻ The answer depends on where you live, which wireless services you use and when you decide to *take the 5G plunge*.

答案取决于你生活在哪里，你使用哪种无线服务，以及你何时决定使用5G。

take the plunge：(尤指深思熟虑后)果断行事，做出决定。例如：We took the plunge and set up our own business.

Exercises

Ⅰ. Answer the following questions according to the passage.

1. What will affect our daily life?
2. What is a big technology shift?
3. Why do we think 5G is more advanced than 4G?
4. What exactly is 5G?

5. How fast will 5G be?

Ⅱ. Translate the following expressions into English or Chinese.

1. 第五代蜂窝网络　　　　6. industrial robots
2. 技术变革　　　　　　　7. artificial intelligence
3. 简称　　　　　　　　　8. technical ground rules
4. 国家竞争力　　　　　　9. exchange data
5. 坦率地说　　　　　　　10. head to

Ⅲ. Fill in the blanks with the proper expressions listed in the box. Change the form if necessary.

| transition | define | noticeable | roughly | designate | artificial | competitive |
| install | upgrade | addict | drone | download | decade | transmission | multiple |

1. _____ speaking, we receive about fifty letters a week on the subject.
2. Are you _____ to your smartphone?
3. Helicopters have been _____ and modernized.
4. Above him an invisible plane _____ through the night sky.
5. The _____ to a multi-party democracy is proving to be difficult.
6. There are efforts under way to _____ the bridge a historic landmark.
7. Users can _____ their material to a desktop PC.
8. The city is dotted with small lakes, natural and _____.
9. British Rail has indeed become more commercial over the past _____.
10. We were unable to _____ what exactly was wrong with him.
11. The car was fitted with automatic _____.
12. It is _____ that women do not have the rivalry that men have.
13. He died of _____ injuries.
14. They had _____ a new phone line in the apartment.
15. Only by keeping down costs will America maintain its _____ advantage over other countries.

Ⅳ. Translate the following sentences into English.

1. 这是我们无线系统十年一次的升级,大约几个月后开始覆盖手机用户。
2. 5G时代的移动互联网速度能让人们在几秒钟内下载完一部电影。
3. 严格地说,5G是一套定义蜂窝网络工作原理的基本技术规则。
4. 答案取决于你生活在哪里,你使用哪种无线服务,以及你何时决定使用5G。
5. 在高质量流媒体视频中,5G的速度尤为明显。

Artificial Intelligence (AI)

Computer scientists have tried to develop techniques that would allow computers to act more like humans since World War Ⅱ. The entire research effort, including decision-making systems, robotic devices, and various approaches to computer speech, is usually called artificial intelligence (AI).

An ultimate goal of AI research is to develop a computer system that can learn concepts (ideas) as well as facts, make common sense decisions, and do some planning.❶ In other words, the goal is to eventually create a "thinking and learning" computer.

A computer program is a set of instructions that enables a computer to process information and solve problems.❷ Most programs are fairly rigid. They tell the computer exactly what to do step by step. AI programs are, however, exceptions to this rule. They can take shortcuts, make choices, search for and try out different solutions, and change their methods of operation.

In many AI programs, facts are arranged to enable the computer to tell how many pieces of information relate to each other and to a given problem. "If/then" rules of reasoning are also programmed to enable the computer to select, organize, and update its information. According to these rules, if something is true, then certain things must follow. Every action makes new possible actions available.

Programs to play chess have been around since the early days of electronic computers, but they tended to be rigid and limited by the skills of the program designers.❸ Detailed instructions on what moves to make and how to respond to an opponent's moves were written into a program. Sometimes the suggestions of several chess experts were included. However, such programs seldom defeated human chess experts. The computer program would tend to be strong in the opening part of the game, but would weaken as the game went on.

Thanks to AI research, all that has changed. Recently, chess-playing computer programs have been developed to defeat most human opponents including chess masters.

Of course, there's more to artificial intelligence than the ability to play games.

Computer scientists are working on dozens of different practical uses for AI programs. These include operating robots, solving math and science problems, understanding speeches, and analyzing images.

Perhaps the biggest use of AI programs is expert advisors for trouble-shooting (locating problems and making repairs) complex systems ranging from diesel engines to nuclear submarines and to the human body. In other words, these AI programs search for trouble, detect and classify problem areas, and give advice.

The use of AI expert advice systems will not be limited to trouble-shooting specific machinery. AI programs are being developed for economic planning, weather forecasting, casting, oil exploration, computer design, and numerous other uses.

AI techniques are also being used to analyze human speeches and synthesize speeches. With the help of laser sensors, AI techniques are being developed to analyze visual information and improve robot capabilities.

Most artificial intelligence systems involve some sort of integrated technologies, for example, the integration of speech synthesis technologies with that of speech recognition. The core idea of AI systems integration is making individual software components, such as speech synthesizers, interoperable with other components, such as common sense knowledge bases, in order to create larger, broader and more capable AI systems. The main methods that have been proposed for integration are message routing, or communication protocols that the software components use to communicate with each other, often through a middleware blackboard system.

New Words and Phrases

intelligence /ɪnˈtelɪdʒəns/	n.	智力，悟性
robotic /rəʊˈbɒtɪk/	a.	机器人的
ultimate /ˈʌltɪmət/	a. 最后的，最终的，根本的；n.	最终
artificial intelligence		人工智能
decision-making system		决策系统
common sense		常识；直觉判断力

Unit 5 Modern Communication

Notes

❶ An ultimate goal of AI research is to develop a computer system *that* can learn concepts(ideas) *as well as* facts, make common sense decisions, and do some planning.

人工智能的最终目标是研发一种可以从概念及事实中学习,做出符合常识的决策并进行某种计划的计算机系统。

"that"作为关系代词引导定语从句,修饰限定先行词"a computer system",并在定语从句中充当主语成分。

as well as:也,还。例如:They sell books as well as newspaper.

❷ A computer program is a set of instructions that *enables* a computer to process information and solve problems.

计算机程序是一组使计算机能够处理信息并解决问题的指令。

enable:使能够。例如:The fall in the value of the pound will enable us to export more goods.

❸ Programs to play chess have been around *since* the early days of electronic computers, but they *tended to* be rigid and limited by the skills of the program designers.

下棋程序在计算机产生的初期就存在,但是程序比较刻板而且受到编程人员能力的限制。

since:从那时以来,后来。它与现在完成时或过去完成时连用,例如:I haven't played rugby since I left university.

tend to:倾向于。例如:It tends to rain here a lot in the spring.

Part 2 Simulated Writing

汇票(Bill of Exchange)

汇票(bill of exchange/draft)是国际贸易结算中使用最多的票据。它是由一方向另一方签发的,要求受票人立即或在一定时间内,对某人或其指定的某人支付一定金额的无条件支付命令书。因此,一张汇票涉及三方:

出票人——开具、签署和提示汇票者;

受票人——汇票已对他开出而尚未给他承兑；

受款人——出票人指示向他付款或签票人承诺向他付款的人。

● Sample 1

下面是一张托收(collection)项下的汇票样例：

No.：663/99

Exchange for £485　　　　Beijing, China, 8th March, 2019

　　At 120 days sight of this FIRST of Exchange (the SECOND of the same tenor and date being unpaid) pay to the order of China National Machinery Import & Export Corporation the sum of Pound sterling four hundred and eighty-five only.

To：West Coast Import Co. Ltd
　　44 Dock Street,
　　Liverpool, England

　　　　　　　　　　　　　　　　　China National Machinery
　　　　　　　　　　　　　　　　　Import & Export Corporation
　　　　　　　　　　　　　　　　　　　　　Manager
(Inv. NO.：663/99)　　　　　　　　　　　　(Signed)

（汇票正本）

Notes

tenor /ˈtenə/	n. 要旨
sterling /ˈstɜːlɪŋ/	n. 英国货币；英镑
to the order of...	按……的指示(表示汇票可背书转让)
at ... days (after) sight	(远期汇票)见票后……天
FIRST/SECOND of Exchange	汇票正本/副本。汇票通常是两张一套，但债务只有一笔，因此正本上注明"副本未付"(the SECOND of the same tenor and date being unpaid)，而副本上则说明"正本未付"(the FIRST of the same tenor and date being unpaid)。若其中一张付讫，则另一张作废。

Sample 2

下面是一张信用证(letter of credit)项下的汇票样例：

NO.:1154/99 Exchange for US＄23 176.00　　　　　Beijing,China,6th May,2019 　　At <u>120 days</u> sight of this SECOND of Exchange (the FIRST of the same tenor and date being unpaid) pay to the order of <u>Bank of China, H. O. B. C.</u>,Beijing the sum of <u>US Dollars twenty-three thousand one hundred and seventy-six only</u> drawn under Documentary Credit No. 2007/48872 of Bank of New South Wales. Value received and charge to account To:<u>Bank of New South Wales,</u> 　　<u>King and Castlesreagh Street,</u> 　　<u>Sydney, NSW, Australia</u> Account of <u>New South Wales</u>　　　　　　　China Textile Machinery 　　　　　<u>Machinery Import Co. Ltd.</u>　　Imp. & Exp. Corp., 　　　　　<u>Sydney, Australia</u>　　　　　　　　Beijing Branch 　　　　　　　　　　　　　　　　　　　　Manager 　　（Inv. No.:1154/99）　　　　　　　　　（singed）

（汇票副本）

Notes

H. O. B. C. (Head Office of Bank of China)	中国银行总行
draw under...	根据……开立
value received and charge to account	指汇票金额全部支取,银行收费记入买方账户

Practice

Ⅰ. Answer the following questions based on the above drafts.

　Sample 1

　1. Who is the buyer?

　2. Is this draft the original one?

　3. What's the total amount of this draft?

Sample 2

1. Who is the seller?

2. Under which credit is this draft drawn?

3. What's the total amount of this draft?

Ⅱ. Complete the following bill of exchange with the given information.

编号:20-0414

发票编号:SA1148

金额:US $52 410.00

付款日期:见票后 60 天

受款人:BANK OF CHINA, H. O. B. C.

出票人:SHENZHEN SHANHE PRINTING EQUIPMENT CO. ,LTD

受票人:BANQUE NATIONALE DE PARIS, PARIS

出票依据:BANQUE NATIONALE, PARIS L/C NO. 20040582

出票日期:2019 年 4 月 2 日

NO.: 1	
Exchange for 2 Beijing,China, 3	
At 4 sight of this SECOND of Exchange (the FIRST of the same tenor and date being unpaid) pay to the order of 5 the sum of 6 drawn under 7 .	
Value received and charge to account	
To: 8	
(Inv. No.: 9)	
	10 (signed)

Part 3   Speaking

Payment

● **Dialogue 1**

A: Now that the price has been settled, let's go on to the terms of payment.

B: As usual practices, we only accept payment by confirmed irrevocable letter of credit.

A: Well, could you make an exception and accept documents against payment, Ms. Hou?
B: Sorry, we couldn't. But after more business between us, maybe we could consider it.
A: To be frank, a letter of credit would increase the cost of our imports. When we open a letter of credit with a bank, we have to pay a deposit. That will tie up our money and add to the cost of what we import.
B: I think you can consult with your bank to see if they will reduce the required deposit to a minimum.
A: It would help us greatly if you would accept D/A or D/P. It makes no difference to you, but it does to us.
B: Mr. Hill, I understand what you mean, but a confirmed irrevocable letter of credit can give the exporter the additional protection of the banker's guarantee. Therefore we always require L/C for our exports.
A: It seems I have to agree with you.
B: That's fine. Thank you.

● Dialogue 2

A: Mr. Zhao, when do we have to open the L/C if we want the goods to be shipped in July?
B: Usually a month before the delivery date your L/C should reach us.
A: Why should such a long time be needed?
B: You know, we should get the goods ready, make out the documents and book the shipping space. All this takes time. You cannot expect us to make delivery in less than a month.
A: Then, I'll arrange for the L/C to be opened as soon as I get home.
B: All right, thank you.
A: Mr. Zhao, how about the validity date of our L/C?
B: Usually the validity date of the L/C should be fixed two weeks after the shipment date.
A: I see. Now I want to know what documents we will have.
B: In usual practice, we provide a full set of shipping documents such as the bill of lading, the invoice, the insurance policy, the certificate of origin and the certificate of inspection.
A: Very good, Mr. Zhao. Would you mind sending us the packing list with the other documents?
B: OK. I'll take it down.
A: Thank you.

● **Dialogue 3**

A: Shall we go on to the terms of payment?

B: Sure. Would you explain your proposition about the terms of payment?

A: As this is your first order, we would like payment prior to delivery.

B: I see your point. But we prefer payment after delivery, because these goods are very expensive.

A: I know, but why does that mean you should pay after delivery?

B: It's a large order, so if we give an advance payment, we will have money trouble, because it will take several months to recover the cost.

A: I understand. You see we have the same problem. In addition, the advance payment is made to us as a general rule.

B: Let's do it this way. We will pay in installments. What do you think of that?

A: That's OK. But how are you going to make your payment?

B: What we'd like to do is make the initial payment of half the total, and pay off the rest in one-month installments after that.

A: That sounds reasonable, but I'll have to check with my manager. Tomorrow I'll let you know our reply.

B: That's fine. I'll be waiting for that.

New Words and Phrases

confirmed /kənˈfɜːmd/	a. 承兑的
validity /vəˈlɪdɪti/	n. 有效性
invoice /ˈɪnvɔɪs/	n. 发票
installment /ɪnˈstɔːlmənt/	n. 分期付款
tie up	冻结
consult with...	与……商量
make out	开具

Notes

terms of payment	付款条件
D/A (documents against acceptance)	承兑交单
D/P (documents against payment)	付款交单
insurance policy	保险单

certificate of origin 产地证明
certificate of inspection 商检证明
advance payment 预付款

Useful Expressions

1. Would you please tell us the terms of payment? 能否告诉我方付款条件?
2. Would you explain your proposition about the terms of payment? 能解释一下您对付款条件的建议吗?
3. An L/C would cause some inconvenience to the importer. 信用证对进口商会造成一定的不便。
4. An L/C would inevitably add to the cost of our imports. 信用证不可避免地会增加我方进口的成本。
5. We hope you will offer us more favorable terms. 我方希望贵方能够提供对我方更有利的条款。
6. It would help us greatly if you would accept D/P. 如果贵方能接受付款交单方式,这必定会极大地有利于我方。
7. We should greatly appreciate it if you could agree to D/P terms. 如果贵方能够同意付款交单,我方将不胜感激。
8. We only accept payment by confirmed and irrevocable letter of credit. 我方仅接受不可撤销信用证的支付方式。
9. Our normal terms of payment are at sight L/C. 我方惯常的支付条款是即期信用证。
10. An irrevocable L/C, as a matter of fact, protects the seller as well as the buyer. 事实上,不可撤销信用证不仅保护买方利益,还保护卖方利益。
11. Your L/C must reach us 30 days before delivery. 贵方的信用证务必于交付前30天到达我方。
12. As the time of shipment is drawing near, we wish you could open the L/C without delay. 随着装运时间的临近,我方希望贵方能尽快开出信用证。
13. For the sake of future business, we'll accept payment in installments this time. 为了将来的合作,这次我方接受分期付款。
14. You can pay the rest in ten monthly installments. 贵方可以将剩余款项按10个月分期付款。

Unit 6
Electric Technology

Part 1 Reading

 **Passage A**

Electric Motors

Each type of motor has its particular field of usefulness. Because of its simplicity, economy, and durability, the induction motor is more widely used for industrial purposes than any other type of AC motor, especially if a high-speed drive is desired. ❶

If AC power is available, all drives requiring constant speed should use squirrel-cage induction motors or synchronous motors because of their ruggedness and lower cost. Drives requiring varying speeds, such as fans, blowers, or pumps, may be driven by wound-rotor induction motors. However, if there are machine tools or other machines requiring adjustable speed or a wide range of speed control, it will probably be desirable to install DC motors on such machines and supply them from the AC system by motor-generator sets or electronic rectifiers.

Almost all constant-speed machines may be driven by AC squirrel-cage motors because these motors are made with a variety of speed and torque characteristics. ❷ When large motors are required or when the power supply is limited, the wound-rotor motor is used, even to drive constant-speed machines.

For varying-speed service, wound-rotor motors with resistance control are used for fans, blowers, and other apparatus for continuous duty and are used for cranes, hoists, and other installations for intermittent duty. The controller and resistors must be properly chosen for the specific application. Synchronous motors may be used for almost any constant-speed drive requiring about 100 hp or over.

Cost is an important factor when more than one type of AC motor is applicable. The squirrel-cage motor is the least expensive AC motor of the three types considered and requires very little control equipment. The wound-rotor is more expensive and requires additional secondary control. The synchronous motor is even more expensive and requires a source of DC excitation, as well as special synchronizing control to apply the DC power on the correct instant. When very large machines are involved, as, for example, 1 000 hp or over, the cost may change considerably and should be checked on an individual basis.

The various types of single-phase AC motors and universal motors are used very little in industrial application, since poly-phase AC or DC power is generally available. When such motors are used, they are usually built into the equipment by the machinery manufacturer, as in portable tools, office machinery, and other equipment.❸ These motors are, as a rule, especially designed for the specific machines with which they are used.

New Words and Phrases

motor /ˈməʊtə/	n. 电动机
durability /ˌdjʊərəˈbɪlɪti/	n. 耐久性,耐用
ruggedness /ˈrʌɡɪdnəs/	n. 坚固性
blower /ˈbləʊə/	n. 鼓风机
rectifier /ˈrektɪfaɪə/	n. 整流器
torque /tɔːk/	n. 转矩
apparatus /ˌæpəˈreɪtəs/	n. 器械,仪器,设备,装置,器官,机构
crane /kreɪn/	n. 起重机
hoist /hɔɪst/	n. 卷扬机
controller /kənˈtrəʊlə/	n. 控制器
resistor /rɪˈzɪstə/	n. 电阻器
excitation /ˌeksaɪˈteɪʃən/	n. 励磁
instant /ˈɪnstənt/	n. 瞬间
induction motor	感应电动机
AC motor	交流电动机
high-speed drive	高速驱动
constant speed	恒速
squirrel-cage induction motor	鼠笼式感应电动机
synchronous motor	同步电动机

wound-rotor induction motor	绕线式转子感应电动机
DC motor	直流电动机
motor-generator sets	电动机-发电机组
hp(horse power)	马力
single-phase	单相
universal motor	交直流两用电动机
poly-phase	多相
as a rule	通常

Notes

❶ Because of its simplicity, economy, and durability, the induction motor is more widely used for industrial purposes than any other type of AC motor, especially if a high-speed drive is desired.

与其他型号的交流电动机相比,感应电动机因其简单、经济和耐用而广泛地用于工业中,特别是需要高速驱动时更是如此。

❷ Almost all constant-speed machines may be driven by AC squirrel-cage motors because these motors are made with a variety of speed and torque characteristics.

因为鼠笼式电动机具有多种速度和转矩特性,几乎所有的恒速机器均可由交流鼠笼式电动机驱动。

❸ When such motors are used, they are usually built into the equipment by the machinery manufacturer, as in portable tools, office machinery, and other equipment.

当使用这类电动机时,机械制造商常把它们安装在设备中,如便携式工具、办公机械及其他设备。

Exercises

Ⅰ. Answer the following questions according to the passage.

1. What are the advantages of the induction motor?

2. Why should all drives requiring constant speed use squirrel-cage induction motors or synchronous motors if AC power is available?

3. When is the induction motor most useful?

4. What kind of motor is used when large motors are required or when the power supply is limited?

5. What factor should be considered when more than one type of AC motor is applicable?

II. Translate the following expressions into English or Chinese.
1. 交流电动机
2. 感应电动机
3. 恒速
4. 单相
5. 持续作业
6. DC motor
7. synchronous motor
8. varying-speed
9. portable tool
10. power supply

III. Translate the following sentences into English.
1. 鼠笼式电动机以其坚固耐用和低成本而著称。
2. 鼠笼式电动机几乎不需要控制装置。
3. 绕线式转子电动机需要次级控制。
4. 几乎任何一种需要 100 马力以上的恒速驱动装置都可以使用同步电动机。
5. 同步电动机需要直流励磁电源和专门的同步控制器。

Passage B

Resistance, Capacitance and Inductance

Resistors, capacitors and inductors form important elements in electronic circuitry. It is essential to know something about resistance, capacitance and inductance.

1. Resistance

Resistance is the opposition to the current and is represented by the letter symbol R. It is the ratio of voltage to current. This relationship between voltage and current, called Ohm's Law, can be stated in an equation
$$i = u/R$$
where R is the resistance in ohms(Ω) if u and i are in volts and amperes, respectively. A larger amount of resistance is commonly expressed in kilo-ohm($k\Omega$) and mega-ohm ($M\Omega$).

Resistors may be classified as fixed or variable in their types and also as linear and nonlinear.

2. Capacitance

A circuit which is able to store electrostatic field energy is said to possess capacitance. The property is defined in terms of the charge stored per unit of potential difference at its terminals, according to the equation

$$q = Cu$$

where C is capacitance, the units of which are farads (symbol F) when u and q are in volts and coulombs, respectively. However, farad is too large a unit to be used in radio calculation, so microfarad (μF) and micro microfarad (pF) are generally used.

Capacitors are used to smooth varying DC supplies by acting as a reservoir of charge. They are also used in filter circuits because capacitors easily pass AC (changing) signals but they block DC (constant) signals.

3. Inductance

A circuit is said to possess inductance if it is able to store magnetic field energy. Winding wire around a suitable mold to form a coil forms an inductor. The property is defined by the relationship

$$u = L \cdot di/dt$$

where L is inductance, the units of which are henries if u and i are in volts and amperes, respectively, with t in seconds. Therefore, 1 V will cause the current to change at the rate of 1 A/s in an inductance of 1 H.

All coils have inductance. Inductance is the property of opposing any change of current flowing through a coil. A small inductor would provide less opposition at the same frequency.

New Words and Phrases

resistance /rɪˈzɪstəns/	n.	电阻
capacitance /kəˈpæsɪtəns/	n.	电容
inductance /ɪnˈdʌktəns/	n.	电感
capacitor /kəˈpæsɪtə/	n.	电容器
inductor /ɪnˈdʌktə/	n.	电感器
voltage /ˈvəʊltɪdʒ/	n.	电压
current /ˈkʌrənt/	n.	电流
equation /ɪˈkweɪʒən/	n.	方程式,等式
ohm /əʊm/	n.	欧姆(电阻单位)
volt /vəʊlt/	n.	伏特
ampere /ˈæmpeə/	n.	安培
respectively /rɪˈspektɪvli/	ad.	各自地,分别地
linear /ˈlɪniə/	a.	线性的,线的
electrostatic /ɪˌlektrəʊˈstætɪk/	a.	静电的

possess /pəˈzes/	v. 具有
property /ˈprɒpəti/	n. 性质,特性
define /dɪˈfaɪn/	v. 为……下定义
term /tɜːm/	n. 术语,专门名词
terminal /ˈtɜːmɪnəl/	n. 终端,接线端
farad /ˈfæræd/	n. 法拉(电容单位)
filter /ˈfɪltə/	n. 滤波器
mold /məʊld/	n. 模子,铸型
henry /ˈhenri/	n. 亨利(电感单位)
coil /kɔɪl/	n. 线圈,绕组
Ohm's Law	欧姆定律
potential difference	电势差
DC(direct current)	直流电
AC(alternating current)	交流电

Part 2 Simulated Writing

保险单(Insurance Policy)

在国际贸易中,货主为了转嫁货物在运输途中的风险,通常要投保货物运输险。如货物一旦发生承保范围内的风险损失,即可从保险公司得到经济上的补偿。

国际货物运输保险属于财产保险的范畴,它以运输过程中的各种货物作为保险标的,被保险人(买方或卖方)向保险人(保险公司)按一定金额投保一定的险别,并交纳保险费。保险人承保以后,如果保险标的在运输过程中发生约定范围内的损失,就按照规定给予被保险人经济上的补偿。

国际货物运输保险种类很多,其中以海上货物运输保险起源最早、历史最久,应用也最广。其保险单上的内容除了文件名称之外,还包括:

(1)保险公司(THE INSUREE);
(2)被保险人(THE INSURED);
(3)保险货物项目(DESCRIPTION OF GOODS);
(4)承保险别(CONDITIONS);
(5)运输船只(PER CONVEYANCE S. S.);
(6)开航日期(SLG/ON OR ABT.);
(7)保险金额(AMOUNT INSURED);
(8)保险费(PREMIUM);

(9)费率(RATE);
(10)赔款偿付地点(CLAIM PAYABLE AT);
(11)出单地点和日期(ISSUING OFFICE & ISSUING DATE);
(12)保险人签字(SIGNED)。

Sample

中国人民保险公司
The People's Insurance Company of China

ORIGINAL	总公司设于北京 Head Office: BEIJING	一九四九年成立 Established in 1949
发票号码 Invoice No.: 20115	保险单 INSURANCE POLICY	号次: NO.: SZ76 200300115

中国人民保险公司(以下简称本公司)
THIS POLICY OF INSURANCE WITNESSES THAT THE PEOPLE'S INSURANCE COMPANY OF CHINA (HEREINAFTER CALLED "THE COMPANY")
该保险单根据天津新港集装箱进出口有限公司
AT THE REQUEST OF Tianjin Xingang Container Import & Export CO., LTD
(以下简称被保险人)的要求,由被保险人向本公司缴付约定的
(HEREINAFTER CALLED "THE INSURED") AND IN CONSIDERATION OF THE AGREED PREMIUM PAID TO THE COMPANY BY THE
保险费,按照本保险单承保险别和背面所载条款与下列
INSURED, UNDERTAKES TO INSURE THE UNDERMENTIONED GOODS IN TRANSPORTATION SUBJECT TO THE CONDITIONS OF THIS
特款承保下述货物运输保险,特立本保险单。
POLICY AS PER THE CLAUSES PRINTED OVERLEAF AND OTHER SPECIAL CLAUSES ATTACHED HEREON.

标记 MARKS & NO.	包装及数量 QUANTITY	保险货物项目 DESCRIPTIONS OF GOODS	保险金额 AMOUNT INSURED
AS PER INVOICE NO. 20115 FCL TOKYO 1-50	1 CASE	MEN'S BICYCLE	USD 10 000.00

总保险金额:
TOTAL AMOUNT INSURED:U. S. Dollars Ten Thousand Only

保费:	费率:	运输船只:
PREMIUM:AS ARRANGED	RATE:AS ARRANGED	PER CONVEYANCE S. S.:NANJING
开航日期:	自:	至:
SLG. ON OR ABT. :MARCH 20,2019	FROM:TIANJIN	TO:TOKYO

承保险别:
CONDITIONS:
Covering All Risks and War Risks as per
Ocean Marine Cargo Clauses and War Risks
Clauses(1/1/1981)of The People's Insurance
Company of China(Abbreviated as C. L. C. All Risks and War Risks Including Risk of Breakage)
(Warehouse to Warehouse Clause is included)

保险货物如遇出险,本公司凭本保险单及其他有关证件给付赔偿。
CLAIMS IF ANY,PAYABLE ON SURRENDER OF THIS POLICY TOGETHER WITH OTHER RELEVANT DOCUMENTS
所保货物如发生本保险单项下负责赔偿的损失或事故,
IN THE EVENT OF ACCIDENT WHEREBY LOSS OR DAMAGE MAY RESUIT IN A CLAIM UNDER THIS POLICY IMMEDIATE NOTICE
应立即通知本公司下述代理人查勘。
APPLYING FOR SURVEY MUST BE GIVEN TO THE COMPANY JS AGENT AS MENTIONED HEREUNDER.

中国人民保险公司天津分公司

偿付地点:NIPPON INSURANCE CO. , TOKYO
CLAIM PAYABLE AT _____   THE PEOPLE'S INSURANCE CO. OF CHINA
日期:MARCH 20,2019   TIANJIN BRANCH

地址:中国天津中山东二路3号
Address:3, Zhongshan Road E. 2. Tianjin, China
Cables:42001 Tianjin

New Words and Phrases

hereinafter /ˌhɪərɪnˈɑːftə/	ad. 以下
case /keɪs/	n. 箱
conveyance /kənˈveɪəns/	n. 运输工具
breakage /ˈbreɪkɪdʒ/	n. 破损
relevant /ˈreləvənt/	a. 有关的
survey /ˈsɜːveɪ/	n. 检验, 调查, 勘察
at the request of	应……要求
the insured	被保险人, 投保人
in consideration of	考虑到
amount insured	保险金额
SLG(sailing)	起航
ABT	about 的缩写
All Risks	综合险
War Risk	战争险

Part 3 Speaking

Advertising and Sales Promotion

● **Dialogue 1**

A: Now, let's have a discussion about future advertising campaigns for our company's products. As you know, in order to guide our campaigns to success in the coming year, we should work out a careful plan.

B: That's a good idea, Mike. I think the first thing to discuss is what media we should use and when we should place advertising in the media.

A: Could we put it in detail, Miss Zhang?

B: OK, very often in a campaign, two or more media are used together. We must be always ready to offer collateral material to people after they have seen ads.

A: You get the point, Miss Zhang! Only in this way can we provide foreign customers with an integral picture of our products.

B: Yes, the purpose of our advertising is to draw foreign customers' interest and keep hold of their attention, so that they may do something in return.

A: What media do you recommend the most then?

B: I think television is much more effective if it doesn't matter for us to pay a little more money.

A: It's worth doing so long as the result is satisfactory. Let's agree on television as the major medium. But when shall we place advertising in this medium?

B: How about the beginning of this coming fall?

A: All right, we'll start at the beginning of the fall.

Dialogue 2

A: Excuse me, but aren't you Mr. Auman, the chief manager of Beijing Office of Sony Group?

B: Yes, I am. I guess you must be the manager from the Design Section of Beijing Advertising Corporation.

A: Exactly.

B: You are the right person I'm expecting. Glad to meet you.

A: Glad to meet you, too. I was told that you need advertising designs for your household appliance products. We can provide you with various designs for your ads in the recommended media.

B: Very good. We need designs for the catalogues of our products for sale, and designs for magazine and newspaper ads as well.

A: Can I see your catalogues?

B: Sure. Here you are.

A: At a minimum estimate, 40 designs are needed.

B: Are they colored or white and black?

A: That depends on you.

B: I'd like them all colored except those on newspaper.

A: All right. Any other requirements?

B: The designs must be novel and appealing to eyes. The printing should be of phototype.

A: OK. Don't worry about that.

B: By the way, how much will you charge for all these designs?

A: Well, with everything included, a rough estimate is around 5 000 yuan.

Dialogue 3

A: What do you have there?

B: One of our new products. It's a tire that will give the longest wear in the world.

A: Really?

B: Yes! We think it's going to be the pride of our company.

A: It sounds interesting. Let me have a look.

B: Thank you. Shall I open the bag now?

A: Of course, carry on.

B: Here it is. We've been testing this tire against all the other brands for over five years now.

A: And what were the results?

B: We found that our tire averaged almost twice as little wear as the other brands.

A: Why did it do so well?

B: Please have a look at this. This is a section of the tire. As you can see, these twin beads are all set in very firmly. They will never break or get loose and cause heat under normal driving conditions.

A: How much do you plan to ask for it?

B: I think we can quote you a price of twenty-eight dollars a tire for the heavy duties. However, since we have a branch office in your city, we can give you a discount on shipping. That way the price will be lower.

New Words and Phrases

campaign /kæmˈpeɪn/	n. 活动,运动；战役
media /ˈmiːdiə/	n. 媒体(单数为 medium)
collateral /kəˈlætərəl/	a. 伴随的,佐证的,附属的
integral /ˈɪntɪɡrəl/	a. 完整的,整体的
phototype /ˈfəʊtəʊtaɪp/	n. 凸版照片
rough /rʌf/	a. 粗略的,大致的
tire /taɪə/	n. 轮胎
bead /biːd/	n. 轮胎边,撑轮圈,轮缘
put it in detail	详细说明
household appliance	家用器具
heavy duty	重负荷,重型

Useful Expressions

1. There are a lot of ways to push sales, for example, by TV ads. 有许多办法促销，例如电视广告。
2. My biggest concern now is business promotion. 我现在最关心的是商业促销。
3. There are a lot of ways to push the sale of new products. 有许多推销新产品的方法。
4. Advertisements on TV will cost a lot of money, I suppose. 我想电视广告要花很多钱。
5. In order to guide our campaigns to success in the new market, we should work out a careful plan. 为了使我们在新市场中获得全面成功，我们应该制订一个周密的计划。
6. Very often in a campaign, two or more media are used together. 在活动中，通常是一起使用两到多个媒体。
7. Now, let's discuss the advertising campaign for our company's products. 现在，让我们讨论一下公司产品的广告活动。
8. We should get to know who are the audiences for the advertising media. 我们应该了解对于广告媒体来说谁是观众。
9. It's worth doing so long as the result is satisfactory. 只要结果令人满意就值得一做。
10. By the way, how much will you charge for all these designs? 顺便问一句，所有的设计要付多少钱？
11. Reports from different markets show that this model has met with a favorable reception in many overseas markets. 不同的市场报告显示这种样式在国外市场很受欢迎。
12. I think you sell direct to shops. 我想你直接卖给商店吧。
13. But, to start with, we'll have more difficulties with new products. 但是，刚开始的时候，我们的新产品可能会有很多困难。
14. After all, you can hardly expect us to finance your sales promotion. 总之，你不能期望我们给你的促销提供资金。
15. You are driving a hard bargain. 你在拼命讨价还价。

Unit 7
Inspection Technology

Part 1 Reading

Passage A

The Features of Transducers

A transducer is a device which converts the quantity being measured into an optical, mechanical, or more commonly—electrical signal. ❶

Transducers are classified according to the transduction principle involved and the form of the measurand, such as capacitance inductance transducers, piezo-electric transducers, magnetoelectric transducers, thermoelectric transducers and photoelectric transducers, etc..

1. Capacitance and Inductance Transducers

Measuring techniques used with capacitive and inductive transducers:

(1) AC excited bridges using differential capacitors inductors;

(2) AC potentiometer for dynamic measurements;

(3) DC circuits to give a voltage proportional to velocity for a capacitor;

(4) frequency-modulation methods, where the change of C or L varies the frequency of an oscillation circuit.

The important feature of capacitive and inductive transducers is the infinite resolution and accuracy ±0.1% of the full scale is quoted. ❷

Typical measurands are displacement, pressure, vibration, sound, and liquid level.

2. Piezo-electric Transducers

When a force is applied across the faces of certain crystal materials, such as natural quartz crystals or barium titanate ceramics, electrical charges of the opposite polarity appear on the faces due to the piezo-electric effect. ❸ Since these materials generate an output charge proportional to the applied force, they are most suitable for measuring force-derived variables such as pressure, load, acceleration and force itself.

The transducer's steady-state response is poor. This can be overcome by using measuring amplifiers with very high input impedances (10^{12} to 10^{14} ohm being typical) known as charge amplifiers, but these make the measuring system increasingly expensive.

3. Magnetoelectric Transducers

These employ the well-known generator principle of a coil moving in a magnetic field and is usually used in the velocity transducer.

Some important features of the magnetoelectric transducer are as follows:

(1) output voltage is proportional to the velocity of input motion;

(2) usually they have a large mass, hence they tend to have low natural frequencies;

(3) high power outputs are available.

4. Thermoelectric Transducers

When two dissimilar metals or alloys are joined together at each end to form a thermocouple and the ends are at different temperatures, an emf will be developed causing a current to flow around the circuit. The magnitude of the emf depends on the material used. This thermoelectric effect is widely used in temperature measurement and control systems.

Although they do give a direct output voltage, this is generally small—in the order of millivolt and often requires amplification.

Advantages of thermocouples include:

(1) temperature at localized points can be determined, because of the small size of the thermocouple;

(2) they are robust, with a wide operating range from $-250 \sim 2\,000$ °C.

5. Photoelectric Transducers

The photoelectric or photovoltaic cell makes use of the photovoltaic effect, which is the production of an emf by radiant emergence—usually light incident on the junction of two dissimilar materials. Light traveling through the transparent layer generates a voltage which is a logarithmic function of light intensity. The device is highly sensitive and has a good frequency response; and, because of its logarithmic relationship of voltage against light, is very suitable for sensing over a wide range of light intensities.

New Words and Phrases

transducer /trænzˈdjuːsə/	n. 传感器
transduction /trænzˈdʌkʃən/	n. 转化
measurand /ˈmeʒərənd/	n. 被测变量，被测物理量
thermoelectric /ˈθɜːməʊɪˈlektrɪk/	a. 热电的
photoelectric /ˌfəʊtəʊɪˈlektrɪk/	a. 光电的
capacitive /kəˈpæsɪtɪv/	a. 电容的
inductive /ɪnˈdʌktɪv/	a. 感应的，电感的
potentiometer /pəˌtenʃiˈɒmɪtə/	n. 电位计
dynamic /daɪˈnæmɪk/	a. 动力的，动态的
proportional /prəˈpɔːʃənl/	a. 比例的，成比例的
infinite /ˈɪnfɪnɪt/	a. 无穷的
quote /kwəʊt/	v. 引用，提供
displacement /dɪsˈpleɪsmənt/	n. 位移
impedance /ɪmˈpiːdns/	n. 阻抗
thermocouple /ˈθɜːməˈkʌpl/	n. 热电偶
transparent /trænsˈpærənt/	a. 透明的
logarithmic /ˌlɒɡəˈrɪðmɪk/	a. 对数的
intensity /ɪnˈtensəti/	n. 强度
piezo-electric transducer	压电式传感器
magnetoelectric transducer	磁电式传感器
thermoelectric transducer	热电式传感器
photoelectric transducer	光电式传感器
photoelectric cell	光电管

Notes

❶ A transducer is a device which converts the quantity being measured into an optical, mechanical, or more commonly—electrical signal.

传感器是一种将被测量转化成光的、机械的或者更平常的电信号的装置。

❷ The important feature of capacitive and inductive transducers is the infinite resolution and accuracy ±0.1% of the full scale is quoted.

电容式和电感式传感器的重要特征是：无限的分辨率，可以精确到满量程的±0.1%。

❸ When a force is applied across the faces of certain crystal materials, such as natural quartz crystals or barium titanate ceramics, electrical charges of the opposite polarity appear on the faces due to the piezo-electric effect.

当对某些晶体材料的表面(如天然石英晶体或钛酸钡陶瓷)施加一个力时,其表面的负极电荷就会在压电效应下出现。

Exercises

Ⅰ. Answer the following questions according to the passage.

1. What is the definition of a transducer?

2. Can you write down some kinds of transducers you are familiar with?

3. What are typical measurands of capacitive or inductive transducers?

4. What crystal materials have the piezo-electric effect?

5. What are the advantages of thermocouples?

Ⅱ. Translate the following expressions into English or Chinese.

1. 位移
2. 热电偶
3. 分辨率
4. 阻抗
5. 光电效应
6. magnetic field
7. magnetoelectric transducer
8. potentiometer
9. steady-state response
10. piezo-electric effect

Ⅲ. Translate the following sentences into English.

1. 压电材料能够产生与施加力成比例的输出电荷。
2. 磁电式传感器利用线圈在磁场中运动产生电流的原理制成。
3. 输出电压与输入运动速度成正比。
4. 热电效应被广泛应用于温度测量与控制系统中。
5. 光电管是利用光电效应的原理制成的。

Passage B

Potentiometers

A transducer that is frequently used to convert mechanical position to an electrical representation of that position is the potentiometer. A potentiometer is a resistor which has a sliding contact or wiper that moves along the resistance element. Potentiometers are widely used in electrical equipment where an adjustable voltage level is desired. A familiar example of the potentiometer use is the volume control of a radio receiver.

A potentiometer is referred to as linear if the resistance per unit length is a constant. In this case, the resistance is proportional to distance. Since the current through the resistance element is uniform, voltage drops are proportional to resistance and so

$$E_0/E_i = R/R_T = \theta/\theta_T$$

or

$$E_0 = E_i \theta/\theta_T$$

thus the voltage from the potentiometer wiper provides a direct measure of the shaft position(Fig. 7-1).

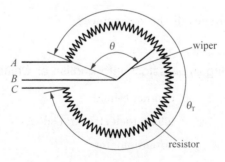

Fig. 7-1 Potentiometer

The most common form of potentiometers has a resistance element made of a length of wire that is wound around a cylindrical core known as a mandrel. Such wire-wound potentiometers can be fabricated in a wide range of resistance values and power ratings. A second basic type is the film potentiometer, which has a resistance element made of a thin film of metal or carbon deposited on a nonconducting base. Since the wiper in a film potentiometer moves along a relatively smooth surface, the torque and wear are generally less than a wire-wound unit. Thus a longer life span and improved reliability are obtained. However, film potentiometers do have the disadvantages of higher cost, lower wattage ratings, and greater temperature sensitivity in terms of resistance changes.

If the resistance element is bent into a circular shape, a single turn potentiometer is obtained. Single turn potentiometers have a usable rotation of slightly less than 360° (although the wiper may have continuous mechanical rotation) since 10° or so are required to provide electrical insulation between the ends of the resistance. Single turn potentiometers are commonly available with diameters between 1/2 and 5 in or more.

As the diameter of a single turn potentiometer is increased, improved accuracy is obtained as a result of the greater length of the resistance element. By shaping the element in the form of a helix and permitting the slider to move along the helix as it rotates, the length can be increased further with a corresponding increase in accuracy. The resultant unit is a multiturn potentiometer, and it is frequently made in 3-, 10- and 15-turn sizes. Obviously, the shaft in a multiturn unit cannot rotate continuously, and mechanical stops are the limits of travel.

New Words and Phrases

wiper /ˈwaɪpə/	n.	滑臂
adjustable /əˈdʒʌstəbəl/	a.	可调的
mandrel /ˈmændrəl/	n.	心轴
film /fɪlm/	n.	薄膜
wattage /ˈwɒtɪdʒ/	n.	瓦特数,功率
helix /ˈhiːlɪks/	n.	螺旋,螺旋线

Part 2 Simulated Writing

面试技巧(Interview Skills)

面试是获得工作机会的一个关键阶段。招聘者往往不只是找一个称职的员工,他们更希望找一个能够进一步发展的人。因此,求职者在面试时的表现就举足轻重了。现将可能帮助面试者取得成功的一些注意事项和常见问答分述如下:

一、面试注意事项

1. 面试前夜要早睡,当日要早起,穿着整齐、干净,而且早餐要吃好。
2. 面试前要思考一下,你在该公司能做什么工作,并随身携带需要的证件、证书等。
3. 考虑一下面试官可能会提的问题,做好回答的准备,同时准备你想问的问题。

4. 看一下你的学业记录,记住你在学校学过哪些课程,哪些同你申请的工作有关,同时要列出你在学校的学习成绩、交往活动与爱好。

5. 了解你所申请的工作,即该职业的具体要求与职责。

6. 要准时到达或适当提前一些时间到达面试场地。

7. 面试时不要紧张,要尽可能给面试者留下自信的印象。

8. 要有礼貌,说话语音清晰、面带微笑,这样做可以缓和气氛。

9. 回答问题要简练、准确,要尽可能和工作联系起来,不要批评你以前的雇主。

10. 对该工作要显得有兴趣,在适当的时候问些问题,对不明白的地方可以坦率地提出。

11. 表现出热情,离开前最好说"I look forward to hearing soon from you about the job"(盼望不久就能听到工作的消息)。

12. 面试结束时,如面试官觉得满意,他可能会把决定告诉你,或者说以后再打电话通知。

二、面试者可能提的问题

1. Do you have any special skills?(你有什么特长吗?)

2. What kind of experience do you have for the job?(你做这工作有什么经验?)

3. What kind of education do you have?(你受过什么教育?)

4. Where was your last job?(你此前的工作在哪里?)

5. Why do you want to leave your present job?(你为什么想辞去现在的工作?)

6. Why did you decide to try our company?(你为什么决定到我们公司寻求工作?)

7. Why did you decide to take a job while still in school?(你为什么尚未毕业就决定找一份工作?)

8. Can you give me one or two references?(你能不能给我们提供一至两个证明人?)

9. What do you hope to become?(你希望今后做什么?)

10. Are you prepared to do probationary year?(你愿意实习一年吗?)

11. How about your paying?(你的工资怎么样?)

12. Do you mind working on the night shift?(上夜班你介意吗?)

三、供求职者参考的内容

1. I have to study during the day, so this job suits me very well.(我必须白天学习,所以这工作对我很合适。)

2. The prospects in my present job are not very good.(我目前工作的前景不太好。)

3. Your company has a good reputation.(贵公司声誉极佳。)

4. I want to get more job satisfaction.(我希望获得更多工作上的满足。)

5. I want to find a company that offers opportunities for advancement and training for its new employees.(我想找一个能提供升职机会并培训新员工的公司。)

6. I hope my being late hasn't inconvenienced you. The delay was unavoidable.（我希望我迟到没给你们带来不便。这次耽搁实在是不得已的。）

7. In today's tough job market, a person has to try every company with an opportunity that matches his or her qualifications and experience.（在如今这样严峻的就业市场上，只要哪家公司有合适的就业机会，就得去试试。）

四、求职者可以提的问题

1. What are the hours?（工作时间怎么样？）
2. Is there overtime?（有加班吗？）
3. What would my duties be?（我的职责是什么？）
4. Is there any opportunity to advance?（有提升的机会吗？）
5. What's the salary?（薪水是多少？）
6. When will I know if I have the job?（我什么时候可以知道我是否能得到这份工作？）

五、面试技巧实例

雇主：李豪贸易公司经理李先生　　　　应征者：林小姐　　　应征职位：公司秘书
下面是他们的对话：

Li：Next applicant, please.（李：下一位请进。）

Lin：Good morning, Mr. Li! I am Ailing Lin.（林：李先生，早上好！我是林爱玲。）

Li：Take a seat, please.（李：请坐。）

Lin：Thank you.（林：谢谢。）

Li：I have read through your documents and would like to ask you some questions if you don't mind.（李：我看过你的资料，不介意的话，我想问你几个问题。）

Lin：OK. I am glad to have the opportunity to meet you.（林：好的。很高兴能有机会见到您。）

Li：What kind of work did you do in your past employment?（李：你过去都做过什么工作？）

Lin：I had three employments in the past. The last one was an assistant to the export manager. In this position, I helped the manager file all his documents and draft replies to some of the letters of enquiry from overseas and remind him about appointments he had promised others.（林：我服务过三家公司，最近一家是出口部经理助理。在这个职位上，我协助经理整理所有文件，草拟回信给海外客户，并提醒他约会相关事宜。）

Li：Are you able to type and what is your speed?（李：你会打字吗？速度如何？）

Lin：Yes. My typing speed is not fast, about 40~45 words per minute, but I seldom make mistakes.（林：会。我打字的速度不太快，大约每分钟40～45个字，但是我很

少打错。)

Li: Very good. Here is a sheet of market report. After the interview, use the typewriter outside, and type exactly the same and hand over to Miss Zhao out there. (李:很好,这是一张市场报告,面试后,用外面的打字机照样打,然后交给外面的赵小姐。)

Lin: Yes, I will. (林:是,我会照办。)

Li: If you are accepted, how much do you expect to be paid? (李:如果被录用,你希望月薪多少?)

Lin: My salary used to be around ¥3 600. I don't mind following the scale of your company. (林:我过去的薪水是3 600元,我不介意遵循贵公司的标准。)

Li: Fine, we will inform you of the result next week. (李:很好,下星期我们会通知你面试结果。)

Lin: Thank you. I hope to receive good news from you. Good-bye, Mr. Li. (林:谢谢,希望能收到您的好消息。再见,李先生。)

Part 3　Speaking

Work and Life

◯ **Dialogue 1**

A: Hey, Mary, what makes you so excited?

B: I've found a part-time job at a McDonald's.

A: Congratulations! You can have hamburgers every day then.

B: No, only at weekends. Now I can earn my own pocket money. What are you planning to do this summer?

A: I will work to earn some money for next semester's tuition.

B: Well, if you want a summer job, it's just about time to start looking for.

A: You know I have been working part-time for a year now. I'm a part-time salesman.

B: Selling something to others is interesting.

A: But it's hard. You have to go from door to door, persuading people to buy your products.

B: I think there is no job in the world that is easy.

A: That's true.

B: Do you think having a part-time job affects your studies?

A: It's hard to say. It depends on how you arrange your time. If you manage your time well, you can just study well.

Unit 7 Inspection Technology

● **Dialogue 2**

A: What do you want to do after graduation, John?

B: I want to be a real estate sales agent.

A: Why? You do not major in business or real estate, after all.

B: For money. You know, money is important for a man.

A: But money isn't everything. I think job satisfaction is more important.

B: Sales work is independent and creative, so I can get satisfaction out of it.

A: Yes, but...

B: What about you, Maria? What plans do you have?

A: I want to be an advertisement designer.

B: Good! You can work as **SOHO**. ("SOHO"是 small office home office 的缩写,译为在家办公)

A: Yeah. My time would be more flexible. I like that.

B: Good luck to you, then.

A: The same to you.

● **Dialogue 3**

A: Today, we have so many sports to enjoy, and I wonder how they were created in the past.

B: Well, as far as I know that many modern sports were introduced or created by the British.

A: Really? I do not even know anything about the history of sports.

B: Well, in Britain, there is widespread participation in sports.

A: They must be very crazy about sports.

B: I don't think they are so crazy about sports, but they do take their free time seriously, so they invented so many sports to entertain themselves in their spare time.

A: What kinds of sports were made by the British?

B: Football, tennis, and horseracing.

A: Is that all?

B: No. There are other sports I cannot remember. Oh, yeah. The home of golf is Scotland because the land there is hilly, and snooker is thought to have been invented by Sir Chamberlain who was English.

A: You surely gave a lesson today. I never imagined that a small country contributed so much to international sports.

B: They did.

● Dialogue 4

A: What kind of preparation should a traveler make?

B: Plan your trip carefully including customs and climate. If you are inexperienced, find a good travel agency or travel with a partner.

A: What kind of dress should a traveler wear?

B: He should wear walking shoes, loose clothing, a sun hat and a pair of sun glasses as well.

A: Any points on eating and drinking?

B: Drink bottle water or boiled water. All meat and seafood must be well-cooked and served while hot. Avoid uncooked vegetables.

A: Problems always appear when packing luggage. Is there any travelers' packing list?

B: There is a list of necessity for travelers: clothes, toothbrush, towel, raincoat or umbrella, walking shoes or boots, flashlight, medicine, maps or guide books, knife, sun glasses, camera and the most important, your passport or ID card.

New Words and Phrases

estate /ɪˈsteɪt/	n.	等级,社会阶层,集团,地产
creative /kriˈeɪtɪv/	a.	有创造力的,创造性的,产生的
widespread /ˈwaɪdspred/	a.	分布广的,流传广的,普遍的
participation /pɑːˌtɪsɪˈpeɪʃn/	n.	参与,参加,分享
hilly /ˈhɪli/	a.	多小山的,丘陵的,陡的
snooker /ˈsnuːkə/	n.	一种落袋撞球游戏;v. 阻挠,挫败
inexperienced /ˌɪnɪkˈspɪəriənst/	a.	没有经验的

Useful Expressions

1. What makes you so excited? 你为什么这么兴奋?
2. Congratulations! 祝贺你啊!
3. Now I can earn my own pocket money. What are you planning to do this summer? 我现在能挣零花钱了,你今年夏天计划做什么?
4. It's just about time to... 正是该做……的时间了。
5. It depends on ... 看……情况;这取决于……
6. What do you want to do after graduation, John? 约翰,毕业后你打算干什么?
7. I major in ... ＝My major is... 我的专业是……,我主修……专业
8. Good luck to you, then. 那祝你好运!
   The same to you. 也祝福你!
9. As far as I know... 据我所知……
10. What kind of preparation should a traveler make? 旅行者应该做什么准备?
11. Any points on eating and drinking? 想吃点什么喝点什么呢?
12. Problems always appear when we are doing... 做……时总会出现问题。

Unit 8
Development of Industrial Technology

Part 1 Reading

▶ **Passage A**

Industrial Robot

An industrial robot is a general-purpose, programmable machine possessing certain humanlike characteristics. The most typical humanlike characteristic of a robot is its arm. This arm, together with the robot's capacity to be programmed, makes it ideally suited to a variety of production tasks, including machine loading, spot welding, spray painting, and assembly. The robot can be programmed to perform a sequence of mechanical motions, and it can repeat that motion sequence over and over until reprogrammed to perform some other jobs.

An industrial robot shares many attributes in common with a numerical control machine tool. The same type of NC technology used to operate machine tools is used to actuate the robot's mechanical arm. The robot is a lighter, more portable piece of equipment than an NC machine tool. The uses of the robot are more general, typically involving the handling of work parts. Also the programming of the robot is different from NC part programming. Traditionally, NC programming has been performed offline with the machine commands being contained in a punched tape. Robot programming has usually been accomplished on-line, with the instructions being retained in the robot's electronic memory. In spite of these differences, there are definite similarities between robots and NC machines in terms of power drive technologies, feedback system, the trend toward computer control and even some of the industrial applications. ❶

The popular concept of a robot has been fashioned by science fictions and movies

Unit 8 Development of Industrial Technology

such as *Star Wars*. These images end to exaggerate the robot's similarity to human anatomy and behavior. The human analogy has sometimes been a troublesome issue in industry. People tend to associate the future use of advanced robots in factories with high unemployment and the conquering of human beings by these machines.

Largely in response to this humanoid conception associated with robots, there have been attempts to develop definitions that reduce the humanlike impact. The Robot Institute of America has developed the following definition:

A robot is a programmable, multi-functional manipulator designed to move materials, parts, tools, or special devices through variable programmed motions for the performance of a variety of tasks.

Attempts have even been made to rename the robot. George Devol, one of the original inventors in robotics technology, called this patent application by the name "programmed article transfer". For many years, the Ford Motor Company used the term "universal transfer device" instead of "robot". Today the term "robot" seems to have become entrenched in the language, together with whatever humanlike characteristics people have attached to the device. ❷

New Words and Phrases

robot /ˈrəʊbɒt/	n.	机器人
programmable /ˈprəʊgræməbəl/	a.	可编程的
capacity /kəˈpæsɪti/	n.	容量,能力
loading /ˈləʊdɪŋ/	n.	上料
mechanical /mɪˈkænɪkəl/	a.	机械的
motion /ˈməʊʃən/	n.	动作,运动
attribute /ˈætrɪbjuːt/	n.	特性
numerical /njuːˈmerɪkəl/	a.	数字的
portable /ˈpɔːtəbəl/	a.	便携式的
command /kəˈmɑːnd/	n.	命令
retain /rɪˈteɪn/	v.	保留
feedback /ˈfiːdbæk/	n.	反馈,反馈信息
trend /trend/	n.	某种倾向
fashion /ˈfæʃən/	v.	塑成,形成
image /ˈɪmɪdʒ/	n.	像,肖像,声誉,心目中的形象或概念
exaggerate /ɪgˈzædʒəreɪt/	v.	夸大

anatomy /əˈnætəmi/	n. 人体解剖,人体结构,解剖学,解剖构造
analogy /əˈnælədʒi/	n. 比拟
issue /ˈɪsjuː/	n. 重要问题;议题;争论的问题
conquer /ˈkɒŋkə/	v. 征服;击败
humanoid /ˈhjuːmənɔɪd/	a. 具有人的特点的
conception /kənˈsepʃn/	n. 概念
manipulator /məˈnɪpjʊleɪtə/	n. 遥控装置
performance /pəˈfɔːməns/	n. 性能
robotics /rəʊˈbɒtɪks/	n. 机器人的应用;机器人学
patent /ˈpeɪtənt/	n. 许可证;专利证书;专利权
universal /ˌjuːnɪˈvɜːsəl/	a. 万能的,通用的
entrench /ɪnˈtrentʃ/	v. 牢固地树立
general-purpose	一般用途的
be suited to	适合于
a variety of	多种的
spray painting	喷漆
a sequence of	一系列的
machine tool	机床
punched tape	穿孔带
attach to	赋予

Notes

❶ *In spite of* these differences, there are definite similarities between robots and NC machines *in terms of* power drive technologies, feedback system, the trend toward computer control and even some of the industrial applications.

尽管存在这些差别,但在电力驱动技术、反馈系统、计算机控制化趋势,甚至在某些工业应用方面,机器人和数控机床之间有着明显的相似之处。

in spite of:介词短语,表示"尽管,虽然"。

in terms of:就……来说,"terms"译为"表达方式,措辞,说法"。

例如:He referred to your work in terms of high praise.

他对你的工作高度赞扬。

❷ Today the term "robot" seems to have become entrenched in the language, *together with* whatever humanlike characteristics people have attached to the device.

如今看来,"机器人"这一名称连同人们所赋予它的人性化特点已经在语言中扎下牢牢的根基。

Unit 8 Development of Industrial Technology

"together with"引出的短语在句中作状语,意思为"连同"。

例如:The new facts, together with other evidence, prove the prisoner's innocence.
这些新的事实连同其他证据已证明在押者无罪。

Exercises

Ⅰ. Answer the following questions according to the passage.

1. What's an industrial robot according to the passage?

2. What is the most typical humanlike characteristic of a robot?

3. In spite of the difference between robots and NC machine tools, how are they similar to each other?

4. Why is the human analogy of a robot sometimes a troublesome issue in industry?

5. Why should the robots be redefined?

Ⅱ. Translate the following expressions into English or Chinese.

1. 电力驱动技术   6. production task
2. 反馈系统   7. spot welding
3. 工业应用   8. mechanical motion
4. 科幻小说   9. offline
5. 机器人技术   10. electronic memory

Ⅲ. Fill in the blanks with the proper expressions listed in the box. Change the form if necessary.

suit to	involve	in terms of	rename	associate with
be different from	in common with	attach to	similarity to	in response to

1. Because of its humanlike arm and the capacity to be programmed, the robot ideally _____ a variety of production tasks.
2. The uses of the robot typically _____ the handing of work parts.
3. Robots and NC machines are much similar _____ power drive technologies.
4. About 20 years ago, people _____ the robots _____ the film *Star Wars*.
5. A robot has many attributes _____ a numerical control machine tool.
6. People _____ whatever humanlike characteristics people have _____ the device.

7. Largely _____ this humanoid conception associated with robots, there have been attempts to develop definitions that reduce the humanlike impact.

8. These images end to exaggerate the robot's _____ human anatomy and behavior.

9. Attempts have even been made to _____ the robot.

10. Also the programming of the robot _____ NC part programming.

Ⅳ. Translate the following sentences into English.

1. 在工业方面,机器人的这种人性化的比拟有时会引起麻烦。
2. 工业机器人是一种具有某些人类特点的可编程的通用机械。
3. 通常,数控编程是脱机完成的,指令载于穿孔带上;而机器人编程常常是联机完成的,指令存于机器人的电子存储器中。
4. 《星球大战》之类的电影和科幻小说促成了现今人们熟悉的有关机器人的概念。
5. 和数控机床相比,机器人是一种质量更轻、更便于携带的设备,其用途也更广泛,主要涉及工件的处理。

Passage B

Mechatronics

Mechatronics is nothing new; it is simply the application of the latest techniques in precision mechanical engineering, control theories, computer science and electronics to the design process to create more functional and adaptable products. This, of course, is something many forward-thinking designers and engineers have been doing for years.❶

As shown in Fig. 8-1, mechatronics is the interdisciplinary fusion (not just a simple mixture) of mechatronics, electronics and information technology. The objective is for engineers to complete development, which is why it is currently so popular with industry.

A famous engineer coined the term "mechatronics" in 1969 to reflect the merging of mechanical and electrical engineering disciplines.❷ Until the early 1980s, mechatronics meant a mechanism that is electrified. In the mid-1980s, mechatronics came to mean engineering that is the boundary between mechanics and electronics. Today, the term encompasses a large array of technologies, many of which have become well-known in their own right. Each technology still has the basic element of the merging of mechanics and electronics but now many also involve much more, particularly software and information technology. For example, many early robots resulting from mechanical and electrical systems became central to mechatronics.

Mechatronics gained legitimacy in academic circles in 1996 with the publication of

Unit 8 Development of Industrial Technology

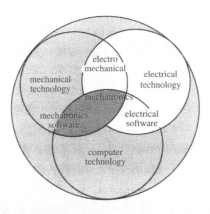

Fig. 8-1 The Interdisciplinary Nature of Mechatronics

the first referred journal: IEEE/ASME Transactions on Mechatronics. ❸ In the premier issue, the authors worked to define mechatronics. After acknowledging that many definitions have circulated, they selected the following for articles to be included in Transactions:❹ "The synergistic integration of mechanical engineering with electronics and intelligent computer control in the design and manufacture of industrial products and processes." The author suggested 11 topics that should fall, at least in part, under the general category of mechatronics:

- modeling and design;
- system integration;
- actuators and sensors;
- intelligent control;
- robotics;
- manufacturing;
- motion control;
- vibration and noise control;
- micro devices and optoelectronics systems;
- automotive systems;
- other applications.

New Words and Phrases

mechatronics /ˌmekəˈtrɔniks/	n. 机电一体化(技术)
interdisciplinary /ˌɪntəˈdɪsɪplɪnəri/	a. 跨学科的,多领域的
fusion /ˈfjuːʒən/	n. 融合,联合,聚变
discipline /ˈdɪsɪplɪn/	n. 纪律,训练(法),学科

encompass /ɪnˈkʌmpəs/	v.	围绕,包含
transaction /trænˈzækʃən/	n.	事项,记录,处理,交易;学报,会刊
synergistic /ˌsɪnəˈdʒɪstɪk/	a.	协作的,合作的
integration /ˌɪntɪˈɡreɪʃən/	n.	整体化,集成,综合
intelligent /ɪnˈtelɪdʒənt/	a.	聪明的,有智力的
category /ˈkætɪɡri/	n.	种类,目录,部门
modeling /ˈmɒdlɪŋ/	n.	建模,造型
actuator /ˈæktʃueɪtə/	n.	调节器,激励器,执行元件
sensor /ˈsensə/	n.	传感器,灵敏元件
optoelectronic /ˌɒptəʊɪlekˈtrɒnɪk/	a.	光电(子)的,光电子学的

Notes

❶ *This*, of course, is *something* many forward-thinking designers and engineers have been doing for years.

当然,这是很多具有超前思想的设计师和工程师多年来一直致力在做的事情。

这里的"this"指代的是上句"to create more functional and adaptable products"。"something"后面省略了 that,不定代词如"something""anything""nothing""everything"等,后面的定语从句只能用 that 引导。

❷ A famous engineer coined the term "mechatronics" in 1969 *to reflect the merging of* mechanical and electrical engineering disciplines.

1969 年,一位著名的工程师杜撰出了这个新词"mechatronics",来反映机械与电气工程学科的融合。

"coin"作动词用,是"杜撰"之意。"to reflect the merging of ..."是不定式作目的状语。

❸ Mechatronics gained legitimacy in academic circles in 1996 *with publication of* the first referred journal:*IEEE/ASME* Transactions on Mechatronics.

1996 年,随着电气电子工程师协会和美国机械工程师协会关于"机电一体化技术"的学报出版发行,机电一体化技术在学术界才得到正式承认。

"with the publication of ..."译为"随着……的出版"。"IEEE/ASME"是电气电子工程师协会和美国机械工程师协会的缩写。

❹ After acknowledging that many definitions have circulated, they selected the following for articles to be included in Transactions:

在承认了许多已流行的关于机电一体化方面的定义后,人们选择了如下一段并收入了学报:

"that many definitions have circulated"是"acknowledging"的宾语从句。

Part 2 Simulated Writing

英文招聘广告(English Employment Ads)

Sample 1

Guangzhou Mike Diesel Engine Co., Ltd.

Guangzhou Mike Diesel Engine Co., Ltd. is a Joint Venture Company established in 1994. Mother companies are Krupp Mike(one of the world's leading companies in medium speed diesel engines) and Guangzhou Diesel Engine Factory. We invite the following candidates in our company:

Project Sales and Services Engineers

—Fluent in spoken and written English and Mandarin
—Be willing to have frequent travel
—Good communication skills, strong technical and commercial sense
—Experience in project sales and services coordination
—University/Polytechnic graduate in power plant or ship diesel engineering

Skilled Workers

We also search specialists for assembly, testing of diesel engines for ships and power plants. Workers shall have good experience or necessary skills for this job.

We offer an attractive salary dependent on qualifications and experience. Interested parties please send your full résumé with the expected salary, photo, contact address and telephone number to Personnel Department. The address of our company:106-1 Fangyuan Road, Guangzhou. Our postcode:510370.

Sample 2

Manager's Assistant (Female)

We are a leading international manufacturer and marketer of premium quality consumer goods based in Europe. We will be establishing our first representative office in Beijing. We are now seeking an experienced versatile female to assist the expatriate manager.

THE RESPONSIBILITIES

1. Assist the setting up and organization of the office.
2. Execute the secretarial duties and office administration.
3. Handle relation and communication with our European head office.
4. Initiate the search for and follow up on possible local partners for cooperation.

THE REQUIREMENTS

1. Minimum 5 years relevant job experience, preferably in multinational companies.
2. Ability to work independently, mature and resourceful.
3. Good command of written and spoken English and Mandarin.
4. Desire to travel.

The right candidate will receive an appropriate and attractive compensation package. Interested candidates please apply in writing in both English and Mandarin as follows:

● Qualifications, job experience with salary history and expected compensation package.

● Recent photo.

And send them to the address below by July 31:

Room 302, International Hotel, IS. ST. Beijing 10008.

All employments will be done through FESCO.

No telephone call will be entertained.

Part 3 Speaking

Work and Life

○ Dialogue 1

A: How far is it to the hotel?

B: About a fifty-minute ride. Is it your first visit to Beijing?

A: Yes, I have been looking forward to seeing the city.

B: There are many scenic spots and historical sites in Beijing, such as the Great Wall,

the Ming Tombs, the Palace Museum, the Summer Palace, and so on.
A: I am really looking forward to visiting them.
B: You must be tired after the long trip. I'm afraid you need a good rest first.
A: That's very kind of you. Oh, that must be Tian'anmen. How big the square is!
B: Yes. It is perhaps one of the largest squares in the world today.
A: That's very interesting. What about the building?
B: It is the Monument to the People's Heroes.
A: I see. By the way, may I know the meeting schedule?
B: Here it is. I'm sure you'll have enough time to see the city. I'll be very glad to show you around.
A: Thank you.
B: Well, here we are. This is the hotel. Let's get off and go to the reception desk.
A: OK.

Dialogue 2

A: Why did you resign your full-time job and start your business at home?
B: In order to look after my parents and children. Also I didn't want to give up my editing work.
A: How did you get started with your home company?
B: I first started working at home using a computer and a fax machine and finally set up my home office.
A: Was it difficult at first?
B: Yes. The cost of the start-up equipment put me in the blue for the first year.
A: How's your business now?
B: I think I am doing well enough to keep it running.

Dialogue 3

A: Good morning, sir! May I have a look at your ticket please?
B: Yes, here it is.
A: Thank you. Will you be travelling with anyone today?
B: No, I am travelling alone.
A: May I see your passport please?
B: Yes, of course. Do you need anything else?
A: Do you have your ID card with you?
B: Yes, here it is. Is it OK if I request a window seat today?
A: That's fine. We'll seat you in Aisle 11 next to the window. Your gate number is 54. Have a good day, and a nice flight.
B: Thank you for your time.
A: It's my pleasure.

Dialogue 4

A: You look a little nervous. Are you OK?

B: Yes, I am fine. I just don't like traveling by airplane very much.

A: Would you like to take my seat near the window?

B: No. It would make me sick.

A: If you don't like to travel, why are you on this flight?

B: I don't like to travel for fun, but I like to travel to see my family.

A: So I take it that you are going to see your relatives.

B: Yes. My three lovely grandchildren are waiting for me in Hong Kong. Why are you travelling today?

A: I like the thrill of travelling. I don't have any grandchildren to visit since I am single.

B: Have you ever been to Hong Kong?

A: No, I haven't.

B: Oh, you are going to love it!

A: I hope so. The airplane ticket to Hong Kong cost me an arm and a leg.

New Words and Phrases

scenic /ˈsiːnɪk/	a. 舞台的,布景的,自然景色的,景色优美的
monument /ˈmɒnjʊmənt/	n. 纪念碑(馆),纪念物,遗址,墓碑
resign /rɪˈzaɪn/	v. 放弃,辞去
edit /ˈedɪt/	v. 编辑,校订,剪辑
aisle /aɪl/	n. (教堂、戏院等的)侧廊,通道,走廊
historical sites	历史遗址(古迹)

Notes

1. ID card:即 identification card,译为"身份证"。

2. window seat,译为"靠窗的座位";aisle seat,译为"靠过道的座位"。

3. I take it that:(口语)以为。
   例如:I take it that they won't accept your proposal.
   我以为他们不会接受你的建议。

4. the thrill of doing sth. = the excitement you get from sth. 做某事获得的兴奋感
   例如:a thrill of joy/horror,译为"一阵喜悦/害怕"。

Unit 8 Development of Industrial Technology

5. an arm and a leg:(口语)昂贵的。

   例如：These resort hotels charge an arm and a leg for a decent meal.

   在这些旅游胜地的餐馆像样地吃一顿要花很多钱。

Useful Expressions

1. How far is it to ...？去……有多远？

2. I am really looking forward to... 我真的希望……

3. That's very kind of you. 你真是太好了。

4. It is perhaps one of the largest squares in the world today. 那可能是目前世界上最大的广场之一。

5. It is the Monument to the People's Heroes. 那是人民英雄纪念碑。

6. Here we are. 我们到了。

7. I first started working at home doing... 我最早在家里做……

8. How's your business now? 你现在的生意如何？

9. I think I am doing well enough to keep it running. 我想我做得挺好,可以使之正常运转。

10. May I have a look at your ticket please? 我可以看一下你的票吗？

11. Do you need anything else? 你还需要别的吗？

12. Thank you for your time. 谢谢你肯花这些时间。

13. So I take it that you are going to see your relatives. 因此我以为你打算看亲戚。

14. Have you ever been to Hong Kong? 你去过香港吗？

Unit 9
NC Technology

Part 1 Reading

▶ Passage A

Machines Using NC

Early machine tools were operated by craftsmen who decided many variables such as speeds, feeds, and depths of cut, etc.. ❶ With the development of science and technology, a new term, Numerical Control(NC) appeared. Controlling a machine tool using a punched tape or stored program is known as Numerical Control. NC has been defined by the Electronic Industries Association(EIA) as "a system in which actions are controlled by the direct insertion of numerical data at some point. The system must automatically interpret at least some portion of these data". ❷

In the past, machine tools were kept as simple as possible in order to keep their costs down. Because of the ever-rising cost of labor, better machine tools, complete with electronic controls, were developed so that industry could produce more and better products at prices that were competitive with those of offshore industries. ❸

NC is being used on all types of machine tools from the simplest to the most complex. The most common machine tools are the single-spindle drilling machine, lathe, milling machine, turning center, and machining center.

1. Single-spindle Drilling Machine

One of the simplest numerically controlled machine tools is the single-spindle drilling machine. Most drilling machines are programmed on three axes:

- The *X*-axis controls the table movement to the right or left.
- The *Y*-axis controls the table movement toward or away from the column.

- The *Z*-axis controls the up or down movement of the spindle to drill holes to depth.

2. Lathe

The engine lathe, one of the most productive machine tools, has been a very efficient means of producing round parts. Most lathes are programmed on two axes:
- The *X*-axis controls the cross motion(in or out) of the cutting tool.
- The *Z*-axis controls the carriage travel toward or away from the headstock.

3. Milling Machine

The milling machine has always been one of the most versatile machine tools used in industry. Operations such as milling, contouring, gear cutting, drilling, boring and reaming are only a few of the many operations that can be performed on a milling machine. ❶ The milling machine can be programmed on three axes:
- The *X*-axis controls the table movement to the right or left.
- The *Y*-axis controls the table movement toward or away from the column.
- The *Z*-axis controls the vertical(up and down) movement of the knee or spindle.

4. Turning Center

Turning Centers were developed in the mid-1960s after studies showed that about 40 percent of all metal cutting operations were performed on lathes. These numerically controlled machines are capable of greater accuracy and higher production rates than the engine lathe. The basic turning center operates on only two axes:
- The *X*-axis controls the cross motion of the turret head.
- The *Z*-axis controls the lengthwise travel(toward or away from the headstock) of the turret head.

5. Machining Center

Machining centers were developed in the 1960s so that a part did not have to be moved from machine to machine in order to perform various operations. These machines greatly increased production rates because more operations could be performed on a work-piece in one setup. There are two main types of machining centers, the horizontal and the vertical spindle types.

(1) The horizontal spindle-machining center operates on three axes:
- The *X*-axis controls the table movement to the right or left.
- The *Y*-axis controls the vertical movement(up and down) of the spindle.
- The *Z*-axis controls the horizontal movement(in or out) of the spindle.

(2) The vertical spindle-machining center operates on three axes:
- The *X*-axis controls the table movement to the right or left.
- The *Y*-axis controls the table movement toward or away from the column.
- The *Z*-axis controls the vertical movement(up and down) of the spindle.

New Words and Phrases

operate /ˈɒpəreɪt/	v. 操作；运行
interpret /ɪnˈtɜːprɪt/	v. 解释；说明
drill /drɪl/	v. 钻孔
milling /ˈmɪlɪŋ/	n. 铣削
turning /ˈtɜːnɪŋ/	n. 车削
table /ˈteɪbəl/	n. 工作台
column /ˈkɒləm/	n. 立柱
carriage /ˈkærɪdʒ/	n. （机床的）滑板；刀架
contouring /ˈkɒntʊərɪŋ/	n. 成形加工
boring /ˈbɔːrɪŋ/	n. 镗孔，镗削加工
reaming /ˈriːmɪŋ/	n. 铰孔
knee /niː/	n. 升降台
accuracy /ˈækjʊrəsi/	n. 精确性，准确度，精度
setup /ˈsetʌp/	n. 安装，设备，机构
horizontal /ˌhɒrɪˈzɒntəl/	a. 水平的
vertical /ˈvɜːtɪkəl/	a. 垂直的，直立的
Numerical Control(NC)	数字控制
Electronic Industries Association(EIA)	（美国）电子工业协会
drilling machine	钻床
turning center	切削中心
machining center	加工中心
engine lathe	卧式车床
cross motion	横向运动
gear cutting	齿轮加工
metal cutting	金属切削
production rate	生产率
turret head	转塔头
lengthwise travel	纵向运动
work-piece	工件

Notes

❶ Early machine tools were operated by craftsmen who decided many variables such as speeds, feeds, and depths of cut, etc..

老式机床通常由工人操作并由他们决定机床速度、进给量、切削深度等参数。

"who decided many variables..."是定语从句，修饰"craftsmen"。

例如：The engineers who we met yesterday have designed a new automatic device.

我们昨天碰到的那些工程师设计出了一种新的自动化装置。

❷ NC has been defined by the Electronic Industries Association(EIA) as "a system in which actions are controlled by the direct insertion of numerical data at some point. The system must automatically interpret at least some portion of these data".

数控被电子工业协会定义为"通过在某些点直接插入数据来控制操作的系统,此系统必须能够自动解释至少这些信息中的一部分"。

该句中"as"的宾语是"a system"，"which"引导的定语从句修饰"prices"。

❸ Because of the ever-rising cost of labor, better machine tools, complete with electronic controls, were developed so that industry could produce more and better products at prices that were competitive with those of offshore industries.

由于劳动成本日益上涨,人们研制出性能更好的配有电控设备的机床,这样企业就能够生产出更多、更好、价格较低的产品,以便和国际上的产品竞争。

"better machine tools"是句子的主语，"were developed"是句子的谓语，"so that"引导结果状语从句，"that"引导的定语从句修饰"prices"。

❹ Operations such as milling, contouring, gear cutting, drilling, boring and reaming are only a few of the many operations that can be performed on a milling machine.

像铣削、成形加工、齿轮加工、钻削、镗削、铰削等只是可以在铣床上进行的一小部分操作。

"that"引导定语从句,修饰"the many operations"。

Exercises

Ⅰ. Answer the following questions according to the passage.

1. What is meant by the term NC?

2. What are the most common machine tools?

3. Has the engine lathe always been a very efficient means of producing round parts?

4. Which machine has always been one of the most versatile machine tools used in industry?

5. How many main types of machining centers are there? What are they?

Ⅱ. Translate the following expressions into English or Chinese.

1. 数字控制   6. turning center
2. 装配有(某设备)   7. machining center
3. 钻床   8. production rate
4. 齿轮加工   9. horizontal movement
5. 穿孔纸带   10. Electronic Industries Association(EIA)

Ⅲ. Fill in the blanks with the proper expressions listed in the box. Change the form if necessary.

affect	lie	lead to	build	set up
print	apply	attract	puzzle	use

1. It was not until the late 1970s that computer-based NC became widely _____.
2. The advantage of this computer _____ in saving the programming time.
3. They _____ an ideal website where he can learn how to write research papers.
4. Why she didn't come still _____ me.
5. Each item in the shop carries a _____ price tag.
6. The professor has answered the question why resistance _____ by temperature.
7. Nowadays NC technology _____ in many machine tools.
8. The building _____; they can move in.
9. The scene on the Huangshan Mountain is really beautiful and _____.
10. The development of NC _____ the development of several other innovations in manufacturing technology.

Ⅳ. Translate the following sentences into English.

1. 随着科学技术的发展,一个新术语——数字控制(NC)诞生了。
2. 过去,人们尽量使机床结构简单,以便降低成本。
3. 从最简单到最复杂的机床都会用到数控技术。
4. 这种数控机床比普通车床具有更高的加工精度和生产率。
5. 由于工件经过一次装配后便能进行多种加工,所以大大提高了生产率。

Passage B

Programming for NC

A program for numerical control consists of a sequence of directions that caused an NC machine to carry out a certain operation, machining being the most commonly used process. Programming for NC may be done by an internal programming department, on the shop floor, or purchased from an outside source. Also, programming may be done manually or with computer assistance.

The program contains instructions and commands. Geometric instructions pertain to relative movements between the tool and the work-piece. Processing instructions pertain to spindle speeds, feeds, tools, and so on. Travel instructions pertain to the type of interpolation and slow or rapid movements of the tools or worktables. Switching commands pertain to on/off position for coolant supplies, spindle rotations, direction of spindle rotations, tool changes, work-piece feedings, clampings, and so on.

1. Manual Programming

Manual part programming consists of first calculating dimensional relationships of the tool, work-piece and worktable based on the engineering drawings of the part, and manufacturing operations to be performed and their sequence. A program sheet is then prepared, which consists of the necessary information to carry out the operation, such as cutting tools, spindle tools, feeds, depths of cut, cutting fluids, power, and tool or work-piece relative positions and movements. Based on this information, the part program is prepared. Usually a paper tape is first prepared for typing out and debugging the program. Depending on how often it is to be used, the tape may be made of more durable mylar.

Someone knowledgeable about the particular process and able to understand, read, and change part programs can do manual programming. Because they are familiar with machine tools and process capabilities, skilled machinists can do manual programming with some training in programming, however, the work is tedious, time consuming, and uneconomical and is used mostly in simple point-to-point applications.

2. Computer-aided Programming

Computer-aided part programming involves special symbolic programming languages that determine the coordinate points of corners, edges, and surfaces of the part. Because numerical control involves the insertion of data concerning work-piece materials and processing parameters, programming must be done by operators or programmers who are knowledgeable about the relevant aspects of the manufacturing processes being used. Before production begins, programs should be verified, either by viewing a simulation of the procession on a CRT screen or by making the parts from an inexpensive material, such as aluminum, wood, or plastic, rather than the material specified for the finished parts.

New Words and Phrases

pertain /pəˈteɪn/	v. 与……直接相关,关于……
interpolation /ɪnˌtɜːpəˈleɪʃən/	n. 插入
clamp /klæmp/	v. 夹住,夹紧
debug /ˌdiːˈbʌɡ/	v. 排除(计算机程序中的)错误
durable /ˈdjʊərəbəl/	a. 耐用的
mylar /ˈmaɪˌlɑː/	n. 聚酯薄膜
parameter /pəˈræmɪtə/	n. 界限;范围;限定因素
verify /ˈverɪfaɪ/	v. 查证,核实
aluminium /ˌæluːˈmɪniəm/	n. 铝
geometric instruction	几何指令
processing instruction	工艺指令
manual part programming	手动零件编程
engineering drawings of the part	零件工程图
manufacturing operation	加工工序
cutting fluids	切削液
paper tape	纸带
point-to-point application	点位加工
symbolic programming language	符号程序语言
processing parameter	工艺参数

Part 2 Simulated Writing

求职信(Application Letter)

　　不用赘言,大家都知道,写求职信的最终目的在于获得职位,不过,现在的公司老板很少是只看信不看人的。一封求职信无论如何文辞并茂、令人心动,公司人事主管不见到求职者本人,是不会给予其工作机会的。因此,求职信的目的在于获得面谈的机会。

一、写求职信的要点

公司的老板大多认为,注重小节的人对重大的事务也会谨慎为之。一个人做人、做事是否谨慎,可以从一封求职信中看出大概来。你别看轻了短短的一封求职信,它可以显露出一个人的爱好、鉴别力、教育程度以及人格特性。下面几个要点便是泄露一个人"机密"的地方,写信人要格外留意。

1. 纸张的选用

最好使用品质优良、白色的信纸,信封要配合信纸的质地和颜色。

2. 书写

求职者若想亲手写信,则字体要写得清晰可辨。如果可能的话,可以把信打印出来,这样看起来比较有商业气息。

3. 格式

信文要适当地排列在信纸中,格式要一致,如采用齐头式(或斜线式)便需全文一致,不可中途改变。

4. 语法、标点和拼写

正确无误的语法、标点和拼写能使读信人感到舒畅,错误的语法或拼写则会给人留下不好的印象。尤其要注意的是,绝不可把收信人的姓名或公司名称写错了。

5. 信封

信封上面的地址要完整,称谓要恰当,信纸的折叠要适当,大小适合信封。

6. 附件

求职信通常不需附加推荐信,除非招聘广告有此要求。遇到这种情形,只需附上复印件即可。求职信内附加邮票或回址信封,强迫对方答复的做法不可取,除非对方有此要求。

二、求职信的内容

求职信的内容通常根据所谋求的工作的性质而定,基本上可以包括下列几项:

1. 写求职信的目的或动机

通常求职信都是针对报纸上的招聘广告而写的,因此,信中需提到何月何日的报纸。有时工作机会是从朋友或介绍所听来的;有时是写信人不知某机构、公司有工作机会,毛遂自荐而写的信。不论是哪一种,求职信上都要说明写信的缘起和目的。

2. 个人资料

写信人应述明自己的年龄或出生年月,教育背景,尤其是和应征的职位有关的训练或教育科目、工作经验或特殊的技能。如无实际经验,则略述在学类似经验亦可。

3. 推荐人

正常的顺序是先获得推荐人同意后再把他们的姓名、地址列入信中,推荐人二至三名即可。

4. 结尾

求职信的结尾希望并请求未来的雇主给予面谈的机会,因此,信中要表明可以面谈的时间。使用的句子要有特点,避免软弱、老生常谈的滥调。

三、机智和良好的判断

写求职信要有机智和良好的判断,下列几点可供参考。

1. 陈述事实,避免表示意见

与训练和经验有关的事实可以陈述出来,但应避免写出这些训练和经验对所应征的工作有怎样的关系或好处。

2. 不要批评他人

如果你要离开现职,可以写出原因但不要用批评的方式,雇主想要了解你,而非你的工作经历。

3. 不要过分渲染自我

你当然认为自己有能力、够资格才会申请这一职位,但不要过分夸大自己的能力或表现得过分自信,尤其不要写出与事实不符的能力或特长来。

4. 留意底薪

有的雇主要你提希望的待遇,你要做出良好的判断,写出你觉得可行的最低薪水。刚开始就业的人应知道,与其寻得一份高薪的工作,倒不如找待遇尚可而有升迁机会的工作。

四、求职信正文

求职信的第一段说明写信的缘起和目的,有些专家认为不宜用分词或从句(如下列各句)作为第一句,因为这类句子被用得太多,显得陈腐,没有突出的特性。

Replying to your advertisement...

Answering your advertisement...

Believing that there is an opportunity...

Thinking that there is a vacancy in your company...

Having read your advertisement...

再比较下列三组句子,其中(1)句较差,经过修改后的(2)句显得较恰当。

1. (1) Replying to your recent advertisement in the Boston Evening Globe, I wish to apply for the position of sales manager...

    (2) In applying for the position of sales manager I offer my qualifications, which I believe will meet your exacting requirements.

2. (1) I believe after reading your advertisement in this morning's Journal that you have just the opportunity I am looking for.

    (2) Your advertisement in this morning's Journal for an adjustment manager prompts me to offer you my qualifications for this position.

3. (1) Having read your advertisement in the "New York Times" for an accountant, I thought you might be interested in my application.

    (2) In your advertisement for an accountant, you indicated that you require the services of a competent person, with thorough training in the field of cost

accounting. Please consider me as an applicant for the position. Here are my reasons for believing I am qualified for this work.

🔍 **注意：**

当求职者不得不提到希望的待遇时，可用下面类似的句子：

1. I hesitate to state a definite salary, but, as long as you have requested me to, I should consider 6 500 yuan a month satisfactory.（我对待遇总是迟迟无法定个确切数目，但既然您要我说明，我认为月薪 6 500 元就满意了。）

2. Although it is difficult for me to say what compensation I should deserve, I should consider... a month a fair initial salary.（虽然我很难说待遇应该是多少，但我认为起薪每月……很合理。）

3. I feel it is presumptuous of me to state what my salary should be. My first consideration is to satisfy you completely. However, while I am serving my apprenticeship, I should consider... a month satisfactory compensation.（我不敢冒昧说出起薪多少。最初我仅想要把工作做好，使您满意。在试用期间，我认为月薪……即可。）

提到待遇时不要过分谦虚或表示歉意，下列句子不宜使用：

1. As for salary, I do not know what to say. Would 4 500 yuan a month be too much?（至于起薪，我不知怎么说，月薪 4 500 元会不会太多?）

2. Do you think I should be asking too much if I said 5 000 yuan a month?（若要求月薪 5 000 元，您会不会觉得太高?）

3. You know what my services are worth better than I do. All I want is a living wage.（对于我工作的价值，您比我更清楚，我仅想够糊口即可。）

关于求职信的结尾用语，应表达出你的自信，语气应不卑不亢：

1.（1）软弱、羞怯的：If you think I can fill the position after you have read my letter, I shall be glad to talk with you.（读完此信后，如您认为我可以填补这个职位的空缺，那么我愿和您一谈。）

（2）改写后：If my application has convinced you of my ability to satisfy you, I should welcome the opportunity to talk with you, so that you may judge my personal qualifications further.（如果我的申请向您证明了我的能力并令您满意，我很乐意有机会与您面谈，那样您可以进一步评判我的个人条件。）

2.（1）怀疑、不妥、不安全的：If you're interested, let me know immediately, as I'm sure an interview will convince you I'm the man for the job.（如贵公司有兴趣，请即告知，我深信与您面谈可以使您相信我适合担任此职。）

（2）改写后：May I have an interview? You can reach me by telephoning at Beijing 010-59139996 between the hours of 7:00～9:00 a.m. and 5:30～9:30 p.m. any evening.（可否面谈? 您可在每天上午七至九时以及下午五时半至九时半致电北京 010-59139996。）

3. 陈腐的句子：Hoping you will give me an interview, I am...（我希望您给予面谈）; Anticipating a favorable decision, I wait for your...（等候您的佳音）; Trusting your

reply will be satisfactory, I remain...（静候满意的答复）。

4.（1）哀求式的句子,不够完整（漏掉面谈时间）：Won't you please give me the chance to interview? I can be reached by calling K-69781.（恳请您给予面谈的机会。给K-69781打电话就能找到我。）

（2）改写后：May I have the opportunity to discuss this matter further with you? My telephone is K-69781. You can reach me between 9:00 and 17:00 during the day.（可否给予面谈以便进一步商讨？我的电话是K-69781,我每天上午九时到下午五时都可接通。）

5.太过自信的句子：I am quite certain that an interview will substantiate my statements. Between two and four every afternoon except Tuesday you can reach me by telephoning 731430.（我深信面谈可以证实我的话,您可在除星期二之外的每天下午二至四时拨打电话731430通知我。）

6.较具体有效的句子：May I have an interview? My residence telephone is（042）4398125. You can reach me by calling that number until June 30. After July 2, my address will be Kent House, Bretton Woods, New Hampshire.（可否给予面谈？我的住处电话是（042）4398125。6月30日以前拨打这个电话都可以找到我。7月2日以后,我的地址为新罕布什尔州布雷顿森林肯特屋。）

五、求职信的语气

求职信要发挥最大的作用,语气必须肯定、自信、有创意而不过分夸张,如能事先洞察雇主的喜好或其人格特点,根据换位思考的原理,配合雇主的特点,则求职者一定可以比其他人占上风,获得面谈的机会。比较下列各句在语气上有何不同：

1. I think that I should probably make a good bookkeeper for you.（我想我会成为贵公司的一名好簿记员。）

2. I am confident that my experience and references will show you that I can fulfill the particular requirements of your bookkeeping position.（我相信我的经验和推荐人可以告诉您,我能够符合贵公司簿记员一职的特定要求。）

3. I have recently completed a course in filing at the Crosby School of Business. I am competent not only to install a filing system that will fulfill the needs of your organization, but I am also well qualified to operate it efficiently.（最近我在克洛斯比商业学校学完了一门档案处理的课程。我不仅可以设置一个符合贵公司要求的档案分类系统,还可以有效地进行操作。）

4. I feel quite certain that as a result of the course in filing which I completed at the Crosby School of Business, I can install and operate efficiently a filing system for your organization.（我相信在克洛斯比商业学校修完这门档案处理课程后,我能够为贵公司设置并有效地操作一个档案分类系统。）

第1句显得语气太弱,写信的人似乎有点羞怯；第2句有特殊风格；第3句太过自信；第4句语气较谦虚。

Part 3 Speaking

Job Interview

(I＝interviewer 面试者   A＝applicant 求职者)

● **Dialogue 1**

I: What kind of character do you think you have?
A: Generally speaking, I am an open-minded person.
I: What is your strongest trait?
A: Cheerfulness and friendliness.
I: How would your friends or colleagues describe you?
A: They say Mr. Sun is a friendly, sensitive, caring and determined person.
I: What personalities do you admire?
A: I admire a person who is honest, flexible and easy-going.
I: How do you get along with others?
A: I get on well with others.

● **Dialogue 2**

I: What kind of person do you think you are?
A: Well, I am always energetic and enthusiastic. That is my strongest personality.
I: What are your strengths and weaknesses?
A: Em, as I have said, I am diligent and industrious. On the other hand, sometimes I'm too hard-working and I put myself under too much pressure to make things perfect.
I: What qualities would you expect of persons working as a team?
A: To work in a team, in my opinion, two characteristics are necessary for a person. That is, the person must be cooperative and aggressive.
I: How do you spend your leisure time?
A: I like playing games and having sports. They are my favorite hobbies.
I: So, what kinds of sports do you like most?
A: Oh, it's hard to narrow it down to just one. I mean, I like all kinds of sports, basketball, swimming, bike riding, and so on. Maybe it is just the reason why I am so energetic and vigorous.

Dialogue 3

I: How would you describe your ideal job?

A: I think the job should make use of the professional experience I have obtained, and offer me opportunity for advancement.

I: Why do you think you might like to work for our company?

A: I feel my background and experience are a good fit for this position and I am very interested. What's more, your company is outstanding in this field.

I: What makes you think you would be a success in this position?

A: My graduate school training combined with my experience as an intern should qualify me for this particular job. I am sure I will be successful.

I: How do you know about this company?

A: Your company is of good repute in this city. I heard much praise to your company.

Dialogue 4

I: I know in your résumé that you have worked in your present company for 3 years. Can you tell me why you want to leave your present job and join us?

A: Because the job I am doing in my present company is of no challenge, but I like challenge. Your firm is a young organization with many innovative ideas. It has been very successful in an expanding market since its establishment 10 years ago. Working for you would be exactly the sort of challenge I am looking for.

I: Why do you think you are qualified for this position?

A: I have excellent communication skills and I am familiar with the procedures for the last company I worked for. Besides, I am a team player and have great interpersonal skills.

New Words and Phrases

trait /treɪt/	n. 品质,性格
cheerfulness /ˈtʃɪəfʊlnɪs/	n. 快乐,高兴
sensitive /ˈsensɪtɪv/	a. 敏感的,容易感受的,神经过敏的,灵敏的
flexible /ˈfleksɪbəl/	a. 柔韧的,柔顺的,灵活的
energetic /ˌenəˈdʒetɪk/	a. 有力的,精力旺盛的,精神饱满的
enthusiastic /ɪnˌθjuːziˈæstɪk/	a. 热情的,热心的,热烈的
cooperative /kəʊˈɒpərətɪv/	a. 合作的,协作的,抱合作态度的;n. 合作团体
aggressive /əˈgresɪv/	a. 侵略的,放肆的,敢作敢为的
leisure /ˈleʒə/	n. 空闲,闲暇;悠闲;a. 空闲的,从容不迫的
narrow /ˈnærəʊ/	a. 狭窄的,范围狭小的,眼光短浅的;v. 变窄,使缩小

vigorous /ˈvɪɡərəs/	a. 朝气蓬勃的,精力充沛的,茁壮的
advancement /ædˈvɑːnsmənt/	n. 前进,进展
intern /ˈɪntɜːn/	vi. 做实习医生;n. 实习医生
repute /rɪˈpjuːt/	n. 名誉,名声,美名,声望
innovative /ˈɪnəveɪtɪv/	a. 革新的,创新的,富有革新精神的
establishment /ɪˈstæblɪʃmənt/	n. 建立,设立,创办,开设;建立的机构(或组织)
generally speaking	总的来说
open-minded	思想开放的
easy-going	平易近人的,容易相处的

Useful Expressions

1. What kind of character do you think you have? 你认为你的性格是什么样的?
2. Generally speaking, I am an open-minded person. 总的说来,我是一个思想开放的人。
3. What is your strongest trait? 你最大的特点是什么?
4. What personalities do you admire? 你欣赏什么性格?
5. How do you get along with others? 你与别人相处得怎么样?
6. I get on well with others. 我和别人相处得很好。
7. What are your strengths and weaknesses? 你的优缺点是什么?
8. How do you spend your leisure time? 你的业余时间怎么度过?
9. What's more... 更重要的是……
10. How do you know about this company? 你是怎么知道这家公司的?
11. I am sure I will be successful. 我相信我会成功的。
12. I heard much praise to your company. 我听到很多对贵公司的称赞。
13. Why do you think you are qualified for this position? 你为什么认为自己能胜任这份工作?

Unit 10
Computer Control

Part 1 Reading

Passage A

CNC

When Numerical Control is performed under computer supervision, it is called Computer Numerical Control(CNC). Computers are the control units of CNC machines. They are built in or linked to the machines via communications channels. When a programmer inputs some information in the program by tape and so on, the computer calculates all necessary data to get the job done.

Today's systems have computer control data, so they are called Computer Numerically Controlled Machines. For both NC and CNC systems, work principles are the same. Only the way in which the execution is controlled is different. Normally, new systems are faster, more powerful, and more versatile units.

CNC machine tools are complex assemblies. However, in general, any CNC machine tool consists of the following units: computers, control systems, drive motors and tool changers.

According to the construction of CNC machine tools, CNC machines work in the following manners:

(1) The CNC machine language, which is a programming language of binary notation used in computers, is not used in CNC machines.

(2) When the operator starts the execution cycle, the computer translates binary codes into electronic pulses that are automatically sent to the machine's power units. The control units compare the number of pulses sent and received.

(3) When the motors receive the pulses, they automatically transform the pulses into rotations that drive the spindle and the lead screw, causing the spindle rotation and slide or table movement. The part on the milling machine table or the tool in the lathe turret is driven to the position specified by the program.

1. Computers

As with all computers, the CNC machine computer works on binary principle using only two characters 1 and 0, for processing the information of precise time impulses from the circuit.❶ There are two states, a state with voltage, 1, and a state without voltage, 0. Series of ones and zeroes are the only states that the computer distinguishes, which is called machine language, and it is the only language the computer understands. When creating the program, the programmer does not care about the machine language.❷ He or she simply uses a list of codes and keys in the meaningful information. Special built-in software compiles the program into the machine language and the machine moves the tool by its servomotors. However, the programmability of the machine is dependent on whether there is a computer in the machine's control. If there is a minicomputer programming, say, a radius(which is a rather simple task), the computer will calculate all the points on the tool path.❸ On the machine without a minicomputer, this may prove to be a tedious task, since the programmer must calculate all the points of intersection on the tool path. Modern CNC machines use 32-bit processors in their computers that allow fast and accurate processing of information.

2. Control Systems

There are two types of control systems on NC/CNC machines: the open loop and the closed loop. The type of control loop used determines the overall accuracy of the machine.

The open-loop control system does not provide positioning feedback to the control unit. The movement pulses are sent out by the control unit and they are received by a special type of servomotor called a stepper motor.❹ The number of pulses that the control sends to the stepper motor controls the amount of the rotation of the motor. The stepper motor then proceeds with the next movement command. Since this control system only counts pulses and cannot identify discrepancies in positioning, the machine will continue this inaccuracy until somebody finds the error.❺

The open-loop control can be used in applications in which there is no change in load conditions, such as the NC drilling machine.❻ The advantage of the open-loop control system is that it is less expensive, since it does not require the additional hardware and electrics needed for positioning feedback. The disadvantage is the difficulty of detecting a positioning error.

In the closed-loop control system, the electronic movement pulses are sent from the control unit to the servomotor, enabling the motor to rotate with each pulse. The movements are detected and counted by a feedback device called a transducer. With each step of movement, a transducer sends a signal back to the control unit, which compares the current position of the driven axis with the programmed position. When the number of pulses sent and received matches, the control unit starts sending out pulses for the next movement.

Closed-loop systems are very accurate. Most have an automatic compensation for error, since the feedback device indicates the error and the control unit makes the necessary adjustments to bring the slide back to the position. They use AC, DC or hydraulic servomotors.

Position measurement in NC machines can be accomplished through direct or indirect methods. In direct measuring systems, a sensing device reads a graduated scale on the machine table or slide for a linear movement. This system is more accurate because the scale is built into the machine and the backlash (the play between two adjacent mating gear teeth) in the mechanisms is not significant.

In indirect measuring systems, rotary encoders or resolves convert a rotary movement to a translation movement. In this system, the backlash can significantly affect measurement accuracy. Position feedback mechanisms utilize various sensors that are based mainly on magnetic and photoelectric principles.

3. Drive Motors

The drive motors control the machine slide movement on NC/CNC equipment. They come in four basic types: stepper motors, DC servomotors, AC servomotors and fluid servomotors.

Stepper motors convert a digital pulse generated by the microcomputer unit (MCU) into a small step rotation. Stepper motors have a certain number of steps that they can travel. The number of pulses that the MCU sends to the stepper motor controls the amount of the rotation of the motor.❾ Stepper motors are mostly used in applications where low torque is required.

Stepper motors are used in open-loop control systems, while AC, DC or hydraulic servomotors are used in closed-loop control systems.

Direct current (DC) servomotors are variable speed motors that rotate in response to the applied voltage. They are used to drive a lead screw and a gear mechanism. DC servomotors provide higher-torque output than stepper motors.

Alternative current (AC) servomotors are controlled by varying the voltage frequency to control speed. They can develop more power than a DC servomotor. They are also used to drive a lead screw and gear mechanism.

Fluid or hydraulic servomotors are also variable speed motors. They are able to produce more power, or more speed in the case of pneumatic motors than electric servomotors. The hydraulic pump provides energy to values that are controlled by the MCU.

4. Tool Changers

Most of the time, several different cutting tools are used to produce a part. The tools must be replaced quickly for the next machining operation. For this reason, the majority of NC/CNC machine tools are equipped with automatic tool changers, such as magazines on machining centers and turrets on turning centers. They allow tools changing without the intervention of the operator. Typically, an automatic tool changer grips the tool in the spindle, pulls it out, and replaces it with another tool.❽ On most machines with automatic tool changers, the turret or magazine can rotate in either direction, forward or reverse.❾

Tool changers may be equipped for either random or sequential selection. In random tool selection, there is no specific pattern of tool selection. On the machining center, when the program calls for the tool, it is automatically indexed into waiting position, where it can be retrieved by the tool-handling device. On the turning center, the turret automatically rotates, bringing the tool into position.

New Words and Phrases

supervision /ˌsuːpəˈvɪʒən/	n.	监督，管理，指导，主管
via /ˈvaɪə, ˈviːə/	prep.	经过，取道，通过某人或某机器等传送某物
rotation /rəʊˈteɪʃən/	n.	旋转
distinguish /dɪˈstɪŋɡwɪʃ/	v.	区别，辨别
servomotor /ˈsɜːvəʊˌməʊtə/	n.	伺服电动机
tedious /ˈtiːdɪəs/	a.	枯燥无味的，冗长的
intersection /ˌɪntəˈsekʃən/	n.	交点，交叉口
discrepancy /dɪsˈkrepənsi/	n.	偏(误)差，不符，不同
compensation /ˌkɒmpənˈseɪʃən/	n.	补偿，补偿物
graduated /ˈɡrædʒueɪtɪd/	a.	标有刻度的
magnetic /mæɡˈnetɪk/	a.	磁的，磁性的
magazine /ˌmæɡəˈziːn/	n.	链式刀库，杂志
retrieve /rɪˈtriːv/	v.	检索
binary notation		二进制符号
execution cycle		执行循环
electronic pulse		电脉冲
lead screw		丝杠，螺杆

milling machine	铣床
stepper motor	步进电动机
proceed with	继续
NC drilling machine	数控钻床
hydraulic servomotor	液压伺服电动机
gear mechanism	齿轮机构
pneumatic motor	气动马达

Notes

❶ *As with* all computers, the CNC machine computer works on binary principle using only two characters 1 and 0, for processing the information of precise time impulses from the circuit.

正像所有计算机那样，CNC 机床上的计算机也只使用 1 和 0 两个字符，按照二进制原理运行，处理来自系统电路的精确时间脉冲信息。

"as with"相当于"as it is the case with"，译为"像……一样"。

❷ When creating the program, the programmer does not care about the machine language.

编程时程序员不必关心机器语言。

when creating the program：现在分词短语作时间状语。

❸ If there is a minicomputer programming, *say*, a radius (which is a rather simple task), the computer will calculate all the points on the tool path.

如果用一个微型计算机进行程序设计，例如半径（一个相当简单的任务），计算机将计算刀具路径上所有的位移点。

"say"为插入语，相当于"for example"。

❹ The movement pulses are **sent out** by the control unit and they are received by a special type of servomotor called a stepper motor.

移动脉冲由控制单元发出，并被一种称为步进电动机的特殊伺服电动机所接收。

send out：发出（光、热、信号等）。

called a stepper motor：过去分词短语作后置定语，修饰"servomotor"。

❺ Since this control system only counts pulses and cannot identify discrepancies in positioning, the machine will continue this inaccuracy until somebody finds the error.

由于这个控制系统只计算脉冲而不能识别位置偏差，因此机床将继续其不准确操作，直到有人发现错误为止。

Unit 10　Computer Control

❻ The open-loop control can be used in applications in which there is no change in load conditions, such as the NC drilling machine.

开环控制适用于载荷状态没有变化的场合,例如数控钻床。

"in which there is no change in load conditions"为定语从句,介词"in"提至"which"前面。

❼ The number of pulses that the MCU sends to the stepper motor controls the amount of the rotation of the motor.

微机单元(MCU)送给步进电动机的脉冲数控制步进电动机的转动角度。

❽ *Typically*, an automatic tool changer grips the tool in the spindle, pulls it out, and replaces it with another tool.

较为典型的是,自动换刀装置会卡紧车床主轴内的刀具,将其拉出,然后换上另一把刀具。

"typically"是评注性副词,修饰整个句子。

❾ On most machines with automatic tool changers, the turret or magazine can rotate in either direction, forward or reverse.

在具有自动换刀装置的大部分机床上,刀架和自动送刀装置可以正向或反向旋转。

Exercises

Ⅰ. Answer the following questions according to the passage.

1. What is the definition of CNC?

2. How many types of control systems are there on NC/CNC machines? What are they?

3. What are the advantages and disadvantages of the open-loop control?

4. What is the advantage of the closed-loop control?

5. What is the advantage of automatic tool changers?

Ⅱ. Translate the following expressions into English or Chinese.

1. 计算机数控　　　　6. DC(direct current)

2. 二进制码　　　　　7. AC(alternative current)

3. 齿轮机构　　　　　　　8. be dependent on

4. 步进电动机　　　　　　9. send out

5. 执行循环　　　　　　　10. be based on

Ⅲ. Fill in the blanks with the proper expressions listed in the box. Change the form if necessary.

| send out | focus on | deny | prove | make sure |
| equip with | shock | accomplish | vary from | call |

1. A week ago he received a notice stating his application was _____.

2. Effective teaching is _____ the learning needs of each student in the class.

3. His mother _____ when she heard about the accident.

4. He _____ a number of e-mail messages to his friends.

5. Position measurement in NC machines can _____ through direct or indirect methods.

6. The actual programming commands needed will also _____ builder to builder.

7. This method _____ to be very successful.

8. _____ that the instructions for the use of this high pressure cooker are strictly observed.

9. The majority of NC/CNC machine tools _____ automatic tool changers, such as magazines on machining centers and turrets on turning centers.

10. The movements are detected and counted by a feedback device _____ a transducer.

Ⅳ. Translate the following sentences into English.

1. 计算机是 CNC 机床的控制单元。它们内嵌于数控机床中或者通过通信渠道与数控机床连接。

2. 新型的数控系统通常速度更快、功率更大、功能更全。

3. 编程时程序员不必关心机器语言，只需要简单地运用一系列代码和符号来表达有用的信息。

4. 这些运动能被一个称为传感器的反馈装置检测并记录下来。每移动一步，传感器就发送一个信号到控制单元，并且将当前驱动轴的位置和程序中的设定位置相比较。

5. 步进电动机常应用在开环控制系统中，而交流、直流或液压伺服电动机常应用在闭环控制系统中。

Passage B

Computer-Aided Design(CAD)

A CAD system is basically a design tool in which the computer is used to analyze various aspects of a designed product. The CAD system supports the design process at all levels—conceptual, preliminary, and final design. The designer can then test the product in various environmental conditions, such as temperature changes, or under different mechanical stresses.

Although CAD systems do not necessarily involve computer graphics, the picture of the object is usually displayed on the surface of a cathode-ray tube(CRT). Computer graphics enables the designer to study the object by rotating it on the computer screen, separating it into segments, enlarging a specific portion of kinematic programs.

Most CAD systems are using interactive graphics systems. Interactive graphics allows the user to interact directly with the computer in order to generate manipulation, and modify graphic displays. Interactive graphics has become a valuable tool, if not a necessary prerequisite, of CAD systems.

The end products of many CAD systems are drawings generated on a plotter interfaced with the computer. One of the most difficult problems in CAD drawings is the elimination of hidden lines. The computer produces the drawing as a wire frame diagram. Since the computer defines the object without regard to one's perspective, it will display all the object's surfaces, regardless of whether they are located on the side facing the viewer or on the back, which normally the eye cannot see.

Various methods are used to generate the drawing of the part on the computer screen. One method is to use a geometric modeling approach, in which fundamental shapes and basic elements are used to build the drawing. The lengths and radii of the elements can be modified. For example, a cylinder is a basic element. The subtraction of a cylinder with a specific radius and length will create holes in the displayed part. Each variation, however, maintains the overall geometry of the part.

Other CAD systems use group technology in the design of parts. Group technology is a method of coding and grouping parts on the basis of similarities in the function or the structure or in the ways they are produced. Application of group technology can enable a company to reduce the number of parts in use and to make the production of parts and their movement in the plant efficient.

Recently CAD systems are using the finite-element method(FEM) of stress analysis. By this approach the object to be analyzed is represented by a model consisting of small elements, each of which has stress and deflection characteristics. The analysis requires the simultaneous solution of many equations. A task, which is performed by the computer, and the deflections of the object, can be displayed on the computer screen by generating animation of the model.

With any of these methods, or others which are used, the CAD system generates on the design stage a single geometric data base which can be used in all phrases of the design and later in the manufacturing, assembling, and inspection processes.

New Words and Phrases

aspect /ˈæspekt/	n.	方面
conceptual /kənˈseptʃuəl/	a.	概念的,构想的
preliminary /prɪˈlɪmɪnəri/	a.	预备的,初步的
segment /ˈsegmənt/	n.	部分,段,节
enlarge /ɪnˈlɑːdʒ/	v.	增大,扩大
kinematic /ˌkɪnɪˈmætɪk/	a.	运动学的,动力学上的
interactive /ˌɪntəˈræktɪv/	a.	相互作用的,相互影响的
interact /ˌɪntəˈrækt/	v.	交流,交往,相互作用
generate /ˈdʒenəreɪt/	v.	产生,创造
manipulation /məˌnɪpjʊˈleɪʃən/	n.	操作
prerequisite /ˌpriːˈrekwɪzɪt/	n.	先决条件,前提;必备条件
plotter /ˈplɒtə/	n.	绘图仪
interface /ˈɪntəfeɪs/	v.	(使)联系,(使)接合;n. 接口,界面
elimination /ɪˌlɪmɪˈneɪʃən/	n.	消除,根除
diagram /ˈdaɪəgræm/	n.	图解,图表
perspective /pəˈspektɪv/	n.	透视,透视画法,透视图
cylinder /ˈsɪlɪndə/	n.	圆筒,圆柱体
simultaneous /ˌsɪməlˈteɪnɪəs/	a.	同时的
computer-aided design(CAD)		计算机辅助设计
cathode-ray tube(CRT)		阴极射线管
group technology(GT)		成组技术
finite-element method(FEM)		有限元法

Part 2 Simulated Writing

求职信(Application Letter)

Sample 1

April 12, 2019

P. O. Box 36
Tsinghua University
Beijing, China 100084

Dear Sir/Madam,

Your advertisement for a Network Maintenance Engineer in the April 10 *Student Daily* interested me because the position that you described sounds exactly like the kind of job I am seeking.

According to the advertisement, your position requires top university, Bachelor or above in Computer Science or equivalent field and being proficient in Windows and Linux system. I feel that I am competent to meet the requirements. I will be graduating from Graduate School of Tsinghua University this year with an M. S. degree. My studies have included courses in computer control and management and I designed a control simulation system developed with Microsoft Visual InterDev and SQL Server.

During my education, I have grasped the principles of my major and skills of practice. Not only have I passed CET-6, but more importantly I can communicate with others freely in English. My ability to write and speak English is out of question.

I would appreciate your time in reviewing my enclosed résumé and if there is any additional information you require, please contact me. I would welcome an opportunity to meet you for a personal interview.

With many thanks,

Wang Lin

Sample 2

April 12, 2019

Room 212, Building 343
Tsinghua University, Beijing 100084

Ms. Yang,

I was referred to you by Mr. Zhang, a partner with your Beijing office, who informed me that the Shanghai office of your company is actively seeking to hire qualified individuals for your Auditor Program.

I have more than two years of accounting experience, including interning as an auditor last year with the Beijing office of ABC company. I will be receiving my MBA this May from Tsinghua University. I am confident that my combination of practical work experience and solid educational experience has prepared me for making an immediate contribution to your company. I understand the level of professionalism and communication required for long-term success in the field. My background and professional approach to business will provide your office with a highly productive auditor upon completion of your development program.

I will be in the Shanghai area within the week of April 16. Please call me at 13600121690 to arrange a convenient time when we may meet to further discuss my background in relation to your needs. I look forward to meeting you soon.

Sincerely,

Cheng Dan

Part 3 Speaking

Job Interview

(I=interviewer 面试者   A=applicant 求职者)

● **Dialogue 1**

I: What kind of work experience do you have?

A: After graduation, I have been working at the Personnel Department of ABC Company.

I: As a telecommunication apparatus company, ABC Company is very different from our trade company.

A: But I deal with people there, the same as what I should do here.

I: You are right. Why are you interested in working in personnel department?

A: I am good with people and have excellent communication skills.

I: Do you consider it a rewarding job?

A: Very much so.

● **Dialogue 2**

I: Can you sell yourself in two minutes? Go for it.

A: With my qualifications and experience, I feel I am hardworking, responsible and diligent in any project I undertake. Your organization could benefit from my analytical and interpersonal skills.

I: Give me a summary of your current job description.

A: I have been working as a computer programmer for five years. To be specific, I do system analysis, trouble shooting and provide software support.

I: Why did you leave your last job?

A: Well, I am hoping to get an offer of a better position. If opportunity knocks, I will take it. I feel I have reached the "glass ceiling" in my current job. I feel there is no opportunity for advancement.

I: How do you rate yourself as a professional?

A: With my strong academic background, I am capable and competent. With my teaching experience, I am confident that I can relate to students very well.

I: What contribution did you make to your current(previous) organization?

A: I have finished three new projects, and I am sure I can apply my experience to this position.

I: What do you think you are worth to us?

A: I feel I can make some positive contributions to your company in the future.

I: What makes you think you would be a success in this position?

A: My graduate school training combined with my internship should qualify me for this particular job. I am sure I will be successful.

I: Are you a multi-tasked individual or do you work well under stress or pressure?

A: Yes, I think so. The trait is needed in my current(previous) position and I know I can handle it well.

Dialogue 3

I: Would you tell me something about yourself?

A: I am 21 years old now and I come from Beijing. My major is accounting and I just graduated from Shanghai University. I received my Bachelor's degree.

I: How were your scores in college?

A: They are all above average 85. For the four years I have studied many subjects, and I worked very hard.

I: Do you think you are proficient in spoken English?

A: I think I speak English quite fluently. I have been attending an English course for oral English for four years. And I often read books and magazines in English.

I: What kind of personality do you think you have?

A: I approach things very enthusiastically. And I don't like to leave things half-done.

Dialogue 4

I: Good morning. What can I do for you?

A: Are you the manager, sir? It's about the Want Ad in this morning's paper. I would like to apply for an assistant sales manager.

I: I see. Won't you sit down and tell me more about you, please?

A: I'm badly in need of a job. Is it possible for me to get the post of sales manager you advertised in this morning's paper?

I: Have you worked anywhere before? You look a little young for the position. What experience have you had?

A: I'm now majoring in economics in university, and I'm looking for a vocational job that would tie with my studies.

I: We are considering the appointment of a new manager. You think a sales manager's job appeals to you while you wouldn't plan to stay here permanently.

A: I worked in a department store in Florida during last Christmas vacation. I think my training and experience have put me in with a chance. I plan to work here until the first of September.

I: I'm afraid that wouldn't qualify you. I'm looking for an elder man, who can be quite fit for the job, familiar with salesmanship, capable of handling personnel and would work with us for a long time.

A: Good day, and thank you anyhow, sir.

New Words and Phrases

telecommunication /ˌtelɪkəmjuːnɪˈkeɪʃən/	n. 电信,(常用复数)电信学
diligent /ˈdɪlɪdʒənt/	a. 勤奋的,用功的,孜孜不倦的
analytical /ˌænəˈlɪtɪkəl/	a. 分析的,分解的,解析的
interpersonal /ˌɪntəˈpɜːsənəl/	a. 人与人之间的,人与人关系的
summary /ˈsʌməri/	a. 概括的,即时的;n. 摘要,概要
description /dɪˈskrɪpʃən/	n. 描写,描述,叙述
specific /spɪˈsɪfɪk/	a. 特有的,特定的,具体的,明确的
academic /ˌækəˈdemɪk/	a. (高等)专科院校的,研究院的,学术的
competent /ˈkɒmpɪtənt/	a. 有能力的,能胜任的,应该做的,足够的,有权力的
confident /ˈkɒnfɪdənt/	a. 确信的,有信心的,过于自信的
contribution /ˌkɒntrɪˈbjuːʃən/	n. 贡献,捐献,捐献物
apply /əˈplaɪ/	n. 应用,运用,使用
internship /ˈɪntɜːnʃɪp/	n. [美]实习医生的职务,实习医生实习期
qualify /ˈkwɒlɪfaɪ/	vt. 使具有资格,使合格
proficient /prəˈfɪʃənt/	a. 熟练的,精通的;n. 能手,专家
personality /ˌpɜːsəˈnælɪti/	n. 人的存在,个性,人格,品格
approach /əˈprəʊtʃ/	vt. 向……靠近;接近;n. 靠近,接近,入门,探讨
enthusiastically /ɪnˌθjuːziˈæstɪkli/	ad. 热情地,热心地,热烈地
vocational /vəʊˈkeɪʃənəl/	a. 行业的,职业的,业务的
permanently /ˈpɜːmənəntli/	ad. 永久地,持久地

Useful Expressions

1. What kind of work experience do you have? 你有什么工作经历？

2. After graduation, I have been working at... 毕业之后,我一直在……工作。

3. Why are you interested in working in... 你为什么喜欢在……工作？

4. I have excellent communication skills. 我的交际能力很强。

5. Do you consider it a rewarding job? 你认为那是一份值得做的工作吗？

6. Can you sell yourself in two minutes? 你可以用两分钟进行自我推荐吗？

7. Give me a summary of your current job description. 对你现有的工作进行概述。

8. I have been working as... 我一直在做……工作。

9. If opportunity knocks, I will take it. 如果机会来了,我会抓住它的。

10. I am hoping to get an offer of a better position. 我一直希望得到一个好的职位。

11. I am confident that I can... 我相信我能……

12. I am sure I can apply my experience to this position. 我相信我可以把经验运用到这个岗位中。

13. Would you tell me something about yourself? 请介绍一下你自己好吗？

14. How were your scores in college? 你在大学的成绩如何？

15. I'm badly in need of a job. 我急需一份工作。

参考文献

[1] 徐存善,周志宇.机电专业英语[M].3版.北京:机械工业出版社,2018.

[2] 曹玲.机电专业英语[M].北京:中国铁道出版社,2014.

[3] 靳敏.机电专业英语[M].北京:机械工业出版社,2018.

[4] 王彤.机电专业英语[M].天津:天津大学出版社,2017.

[5]《机电行业英语》编写组.机电行业英语[M].北京:高等教育出版社,2016.

附　录

附录1　翻译技巧

一、专业英语词汇的特点

专业英语词汇是用来记录和表述各种现象、过程、特性、关系、状态等不同名称的，也称为术语（term）。它在通用的词汇中虽占少数，但举足轻重，往往是段落文章中所要论述的中心词语，如 microprocessor（微处理器）、welding（焊接）、NC（Numerical Control 数控）等。专业英语词汇的主要特征如下：

（1）涌现性：新学科的诞生、新化合物的合成和新物种的发现等，使科技术语随着技术的发展而不断涌现。

（2）单义性：对于某一特定专业或分支，其词义狭窄、形态单一，定义时尽可能避免同形异义或同义异形现象。

（3）中性：术语只有概念意义，没有任何添加色彩，如将"cat"转化为"吊锚"、将"dog"转化为"卡爪器"后，就失去原来猫和狗的形象及人们在用词形式上可能反映的好恶。

（4）国际性：英语和其他印欧语系语言中的部分技术词汇和半技术词汇来源于拉丁语和希腊语，且词汇的专业性越强，在印欧语系语言中的同形同义词越多，国际性也越强。另外，新创造的科技词汇只要在一国使用和流行，英语国家和别的拼音文字国家就可以按照语音对应规律和拼写体系将其转化过来，使之成为自己的文字。例如，transistor（晶体管）一词在英语、德语和法语中是完全一样的。

二、构词法

目前，各行各业都有一些本专业学科的专业词汇，从语言学的角度看，这些专业词汇主要是通过沿用生活中的词汇，或借用外来语，或通过其他方式构成。

（一）派生词

派生词是英语构词法的一个主要手段，是在已有词汇的单词或词干的前面或后面通过加词缀的方法来构成新词。

1. 前缀

英语的前缀通常有固定的含义，添加在单词或词干之前。采用前缀的单词在技术词汇中占有很大比例。

（1）multi-（多）

例如：multimedia（多媒体），multi-sensor（多传感器），multi-processor（多处理器），

multi-protocol(多协议),multi-program(多道程序),multi-state(多态性)。

(2)hyper-,super-(超级)

例如:hypermedia(超媒体),hypertext(超文本),hypercard(超级卡片),hyperswitch(超级交换机),hypercube(超立方),superhighway(超级公路),supermarket(超级市场),superconductor(超导体),super-pipeline(超流水线),superscalar(超标),superclass(超类)。

(3)inter-(相互间,在……间)

例如:interface(接口,界面),interlace(隔行扫描),interlock(连锁),Internet(互联网络),interaction(相互作用,互感),interconnection(互联)。

(4)micro-(微型)/macro-(宏大)

例如:microcode(微代码),microchannel(微通道),microprocessor(微处理器),micromotor(微型电动机),microfilm(微型胶片),Microsoft(微软)。

(5)tele-(远程的,电的)

例如:telemarketing(电话购物),teleconference(远程会议),telephone(电话),telecom(电信),telegraph(电报机),tele-immersion(远程沉浸)。

(6)by-(边,侧,偏,副,非正式)

例如:by-product(副产品),by-pass(旁通),by-road(小路)。

(7)centi-(百分之一)

例如:centimeter(厘米),centigrade(百分度),centigram(厘克),centisecond(厘秒),centigrade thermometer(百分度温度计)。

(8)co-(共同,相互)

例如:cooperation(合作),co-run(共同管理),coexist(共存),collaboration(合作)。

(9)counter-(反)

例如:counteraction(反作用),counterclockwise(逆时针方向),countermeasure(对策,措施),counter-current(反向电流),counterclaim(反索赔),counterbore(沉孔),countercharge(反指控)。

(10)dis-(否定,相反)

例如:disorder(混乱),discharge(放电),disappear(消失),disassembly(拆卸),disconnection(切断,中断)。

(11)en-(使……)

例如:enlarge(扩大),enclose(封闭),encode(编码),enforce(实行,实施,强制),endanger(使危险)。

(12)ex-(出自,向外)

例如:export(出口),extract(抽出,提取),exterior(外部,外面),exchange(交换)。

(13) in-, im-, il-(不,非,无)

例如:incorrect(不正确的,错的),inseparable(不可分的),independency(独立),invisible(看不见的),impossible(不可能的),illogical(不合逻辑的)。

(14) kilo-(千)

例如:kilogram(千克),kilometer(千米,公里),kilowatt(千瓦),kilo-Newton(千牛)。

(15) milli-(毫,千分之一)

例如:millivolt(毫伏),milliliter(毫升),milligram(毫克)。

(16) mis-(误,错,坏)

例如:misfortune(不幸),mislead(误导),misunderstand(误解),misuse(滥用,误用)。

(17) a-(无,非,不)

例如:aperiodic(非周期的),acentric(无中心的)。

(18) anti-(反对,相反,防止,防治)

例如:anti-virus(防病毒),antimagnetic(防磁的),antiwar(反战的),antibody(抗体)。

(19) auto-(自动,自身)

例如:auto-alarm(自动报警器),autorotation(自动旋转),autobiography(自传),automat(自动售货机)。

(20) bi-(两,二,重)

例如:bi-colored(双色的),bimonthly(两月一次的,双月刊),bilingual(双语的),bicycle(自行车),bi-wheel driven(两轮驱动的)。

(21) semi-, hemi-(相当于 half)

例如:semiconductor(半导体),hemicycle(半圆形),semifinal(半决赛),semitone(半音)。

(22) mono-(单,一)

例如:monoplane(单翼飞机),monorail(单轨铁道),monotone(单调,单音),monograph(专著,专论),monopoly(独有权,垄断)。

(23) non-(非,不,无)

例如:nonstop(直达,中途不停),nonmetal(非金属),nonconductor(绝缘体),nonstandard(非标准的,不规范的)。

(24) over-(过分,在……上面,超过,压倒,额外)

例如:overproduction(生产过剩),overload(过载),overcharge(要价太高),overhead(在头顶上的,在上头的),overrun(溢出,超越)。

(25) poly-(多,复,聚)

例如:poly-crystal(多晶体),poly-technical(多工艺的),poly-atomic(多原子的),polygon(多边形),polysyllable(多音节词)。

(26)post-(后)

例如：postwar（战后的），post-liberation（新中国成立后），postdate（在日期后），postgraduate（研究生）。

(27)pre-(预先,在前)

例如：preheat（预热），preschool（学前），predetermine（先定,预定），precondition（前提,先决条件），prepay（预付,提前付）。

(28)re-(再,重新)

例如：rerun（重新运行），rewrite（改写），reset（重新设置），restate（重述），reprint（重新打印）。

2．后缀

后缀是在单词或词干的后部添加词缀。后缀一般只改变词性，词的含义基本不变。

(1)常见构成名词的后缀

①-age(场所,费用,行为或行为的结果、状态、情况)

例如：mileage（英里数），passage（通道），postage（邮费），breakage（破碎,破损），wastage（耗损），shortage（不足,短缺），advantage（利益），usage（使用）。

②-ance,-ence(性质,状况,行动,过程)

例如：abundance（丰富,充裕），ignorance（忽视），intelligence（智力），interference（干扰,干涉）。

③-ant,-ent(……者)

例如：assistant（助手），participant（参加者），agent（代理人），client（委托人,当事人），attendant（服务员,随从人员）。

④-er,-or(……者,……物,……的机械)

例如：capacitor（电容器），contractor（压缩机），distributor（配电盘,燃料分配器），propeller（推进器），cooker（厨具），washer（洗衣机）。

⑤-ese(……的人,……派的,……特色)

例如：Chinese（中国人），Japanese（日本人）。

⑥-ess(女性)

例如：actress（女演员），hostess（女主人），waitress（女服务员），princess（公主）。

⑦-graph(记录仪器)

例如：barograph（气压记录仪），telegraph（电报），spectrograph（分光摄像仪）。

⑧-ing(动作,动作的结果,与某一动作有关)

例如：engineering（工程,工程学），feeling（感觉），greeting（问候），fishing（钓鱼），finding（发现的东西），reading（读数）。

⑨-ism(……主义,宗教,行为,……学,……论,……法,具有某种特性、情况、状态)

例如：capitalism（资本主义），heroism（英雄主义），magnetism（磁力学），hypnotism（催眠术），atomism（原子论）。

⑩-ist(某种主义者或信仰者,从事某种职业或研究的人)

例如：socialist(社会主义者),typist(打字员),scientist(科学家)。

⑪-ivity(性质,情况,状态)

例如：activity(活动性,活动),productivity(生产力,生产率),sensitivity(敏感性,灵敏度),passivity(被动性)。

⑫-ment(行为,状态,过程)

例如：development(发展),agreement(同意,协议),equipment(设备),investment(投资),requirement(需要)。

⑬-meter(计量仪器)

例如：barometer(气压标),telemeter(测距仪),spectrometer(分光仪)。

⑭-ness(性质,情况,状态)

例如：illness(病,疾病),firmness(结实,坚定),idleness(懒惰),business(事物,商业)。

⑮-scope(探测仪器)

例如：telescope(望远镜),spectroscope(分光镜)。

⑯-ship(情况,技能,身份,职位)

例如：friendship(友谊,友好),membership(成员资格),professorship(教授身份),ownership(所有权,所有制)。

⑰-th(动作,过程,状态,性质)

例如：growth(生长,增长),strength(力量),depth(深度)。

⑱-tion(行为的过程、结果、状况)

例如：option(选择,选择权),function(作用),addition(增加),elimination(消灭,排除),execution(完成,执行),situation(位置,处境)。

⑲-ty(性质,状态,情况,构成抽象名词)

例如：safety(安全),beauty(美丽),cruelty(残忍),liberty(自由)。

⑳-ure(行为,行为的结果、状态、情况)

例如：culture(文化教养),pressure(压力),flexure(弯曲)。

㉑-ware(件,部件)

例如：hardware(硬件),groupware(组件),freeware(赠件)。

㉒-y(性质,状态,行为,构成抽象名词及学术名)

例如：difficulty(困难),discovery(发明,发现),possibility(可能性),philosophy(哲学)。

(2)常见构成形容词的后缀

①-able(能……的,可……的,易于……的)

例如：usable(可用的),audible(听得见的),soluble(可溶的)。

②-al(属于……的,与……有关的)

例如：digital(数字的),decimal(小数的,十进位的),elemental(基本的),external(外部的)。

③-ant(……的,具有……性质的)

例如:abundant(丰富的),assistant(辅助的),vacant(空的),ascendant(上升的)。

④-ar(有……性质的,如……的,属于……的,有……的)

例如:familiar(熟悉的),similar(同样的),solar(太阳的),nuclear(原子核的)。

⑤-ary(与……有关的,属于……的)

例如:ordinary(普通的),necessary(必需的),customary(习惯的),elementary(初级的)。

⑥-ed(有……的,如……的)

例如:colored(彩色的),skilled(熟练的),complicated(复杂的)。

⑦-en(由……制成的,含有……质的,似……的)

例如:woolen(羊毛制的),wooden(木制的),golden(金制的),leaden(铅制的)。

⑧-ent(具有……质的,关于……的,有……行为倾向的)

例如:apparent(明显的),frequent(频繁的),patent(特许的)。

⑨-fill(充满……的,具有……性质的,可……的)

例如:useful(有用的),hopeful(富有希望的),powerful(有力的),wonderful(奇妙的)。

⑩-ible(易于……的,可……的)

例如:compatible(兼容的),visible(看得见的),flexible(易弯曲的),sensible(感知的)。

⑪-ic(与……有关的,属于……的,有……特性的)

例如:academic(学术的),elastic(弹性的,灵活的),atomic(原子的),periodic(周期的)。

⑫-ive(有……性质的,属于……的,与……有关的,有……倾向的)

例如:attractive(有吸引力的),active(积极的),expensive(昂贵的),productive(生产的)。

⑬-less(没有,无,不)

例如:useless(无用的),harmless(无害的),careless(粗心的),wireless(无线的)。

⑭-ly(如……的,有……性质的)

例如:friendly(友好的),lovely(可爱的),early(早的),timely(及时的)。

⑮-ous(多……的,有……的)

例如:dangerous(危险的),famous(著名的),poisonous(有毒的),enormous(巨大的)。

⑯-some(充满……的,易于……的,产生……的)

例如:tiresome(令人厌倦的),handsome(漂亮的),troublesome(令人烦恼的)。

⑰-y(多……的,有……的,有点……的)

例如:rainy(有雨的),ready(准备好的),holy(神圣的),empty(空的),tricky(难处理的),noisy(嘈杂的)。

(3)常见构成动词的后缀

①-ate(成为……,处理,使,作用)

例如:eliminate(排除,消灭),circulate(循环,流通),terminate(终止),estimate(估计)。

②-en(使成为,变成,引起)

例如:widen(加宽),darken(使变暗,变黑),strengthen(加强),weaken(削弱)。

③-fy(弄成,使……化)

例如:simplify(简化),specify(指定),testify(证明),beautify(美化)。

④-ze(变成,……化,实行)

例如:characterize(表示……的特征),liquidize(液化),industrialize(使工业化),optimize(完善)。

(4)常见构成副词的后缀

①-ly(每一次,……地)

例如:yearly(每年),daily(每日),greatly(大大地),quickly(迅速地),carefully(认真地)。

②-wards(向……)

例如:outwards(向外),eastwards(向东),forwards(向前),upwards(向上)。

③-wise(方向,方式,状态)

例如:clockwise(顺时针方向),likewise(同样地),otherwise(否则),lengthwise(纵长地)。

(二)复合词

复合词是由两个独立的词合成的一个词,专业技术词汇常由两个或更多的词组成,可构成复合名词、复合形容词、复合动词等,连字符可连接于各词之间,也可省略,还可独立书写。

1. 复合名词

(1)名词+名词

例如:metal+work→metalwork(金属制品),wave+length→wavelength(波长),liquid+crystal→liquid crystal(液晶)。

(2)名词+动名词

例如:machine+building→machine-building(机器制造),book+learning→book-learning(书本知识),hand+writing→handwriting(笔迹,字迹)。

(3)动名词+名词

例如:waiting+room→waiting room(候车室),building+material→building material(建筑材料),swimming+pool→swimming pool(游泳池)。

(4)形容词+名词

例如:short+hand→shorthand(速记),hard+ware→hardware(硬件),soft+ware→

software(软件),black＋board→blackboard(黑板)。

(5)动词＋名词

例如:pick＋pocket→pickpocket(小偷),break＋water→breakwater(防水堤)。

(6)副词＋动词

例如:in＋put→input(输入),out＋put→output(输出,产量),out＋come→outcome(结果)。

(7)动词＋副词

例如:feed＋back→feedback(反馈),get＋together→get-together(联欢会),stand＋still→standstill(停顿)。

2. 复合形容词

(1)名词＋现在分词

例如:peace＋loving→peace-loving(热爱和平的),epoch＋making→epoch-making(划时代的)。

(2)名词＋过去分词

例如:man＋made→man-made(人造的),hand＋made→handmade(手工制造的)。

(3)形容词＋现在分词

例如:good＋looking→good-looking(好看的),fine＋looking→fine-looking(美观的),deep＋going→deep-going(深入的)。

(4)形容词＋过去分词

例如:ready＋made→ready-made(现成的)。

(5)形容词＋名词

例如,new＋type→new-type(新型的),large＋scale→large-scale(大规模的)。

(6)形容词＋名词＋ed

例如:medium＋sized→medium-sized(中型的),noble＋minded→noble-minded(高尚的)。

(7)形容词＋形容词

例如:red＋hot→red-hot(炽热的),light＋blue→lightblue(淡蓝的),dark＋green→darkgreen(深绿的)。

(8)数词＋名词

例如:first＋class→first-class(第一流的),three＋way→three-way(三通的,三项的)。

(9)数词＋名词＋ed

例如:four＋cornered→four-cornered(有四个角的),one＋sided→one-sided(单面的,片面的)。

(10)副词+现在分词

例如:hard+working→hard-working(勤劳的),ever+increasing→ever-increasing(不断增加的)。

(11)副词+过去分词

例如:well+known→well-known(著名的),newly+built→newly-built(新建的),above+mentioned→above-mentioned(上述的)。

(12)介词或副词+名词

例如:under+ground→underground(地下的),off+hand→offhand(即刻的)。

3. 复合动词

(1)名词+动词

例如:work+harden→to work-harden(加工硬化),heat+treat→to heat-treat(热处理),trial+produce→to trial-produce(试制)。

(2)形容词+动词

例如:safe+guard→to safeguard(保卫),white+wash→to whitewash(粉刷)。

(3)介词或副词+动词

例如:over+heat→to overheat(过热),over+write→to overwrite(覆盖),up+set→to upset(推翻),under+line→to underline(在……下面画线)。

(三)转化法

英语中的单词可以从一种词类转化为另一种词类。如名词、形容词、副词、介词可以转化为动词,动词、形容词、副词、介词可以转化为名词。最常见的是名词转化为动词和动词转化为名词。转化后的词义往往与原词义有密切的关系。

1. 名词转化为动词

例如:machine(机器)→to machine(机器加工),time(时间)→to time(计时、定时),format(格式)→to format(格式化)。

2. 动词转化为名词

例如:to talk(交谈)→talk(谈话),to test(测验)→test(检测),to use(使用)→use(用途),to increase(增加)→increase(增加)。

3. 形容词转化为名词

例如:mineral(矿物的)→mineral(矿物质),good(好的)→good(益处),final(最后的)→final(决赛)。

(四)缩略词

词汇缩略是指将较长的单词取其首部或主干构成与原词同义的短单词,或者将组成词汇短语的各个单词的首字母拼接为一个大写字母的字符串。词汇缩略通常在科技文章中被频繁采用。缩略词汇的使用方便了印刷、书写、速记和交流等,但也增加了阅读和理解的难度。

1. 首字母缩略词(acronym)

首字母缩略词由词组中每个单词的第一个字母构成。例如：

DBMS—Data Base Management System(数据库管理系统)

CGA—Color Graphics Adapter(彩色图形适配器)

CAD—Computer Aided Design(计算机辅助设计)

CAE—Computer Aided Engineering(计算机辅助工程)

CAM—Computer Aided Manufacture(计算机辅助制造)

NC—Numerical Control(数控)

2. 截短词(dipped word)

截短词由截取的较长单词的一部分构成，通常截取单词的首部、尾部或首尾部等。例如：telecommunication→telex(电信), memorandum→memo(备忘录), telephone→phone(电话), refrigerator→fridge(冰箱)。

3. 缩写(abbreviation)

缩写即用字符代替单词或词组。例如：e. g. →for example(例如), Ltd→limited(有限的), sq. →square(平方)。

(五)拼缀词

拼缀词以原有的两个或两个以上的词为基础，经过首尾截取或保留其中一个原词，重新组合而成。例如：cermet → ceramics metal(金属陶瓷), comsat → communication satellite(通信卫星), firex → fire extinguishing equipment(灭火设备), insullac → insulation lacquer(绝缘漆), telecamera→television camera(电视摄像机)。

三、词义的引申

英汉两种语言在表达方式上差别很大，在翻译科技文章中的词汇时，若勉强按词典中的释义逐词死译，会使译文生硬晦涩，甚至会造成误解。所以，要在理解原文词义的基础上，根据上下文的逻辑关系和汉语的表达习惯，对词义进行引申，对于专业方面的内容，必须选用专业常用语。引申后的词义虽与词典释义略有不同，但能更确切地表达原文意思。

例如：

(1)There is no physical contact between tool and workpiece.

　　欠佳译文：在工具和工件之间没有有形的接触。

　　引申译文：工具和工件不直接接触。

(2)Public opinion is demanding more and more urgently that something must be done about noise.

　　欠佳译文：公众舆论越来越强烈地要求为消除噪声做某些事情。

　　引申译文：公众舆论越来越强烈地要求管一管噪声问题。

(3)There is a wide area of performance duplication between numerical control and automatics.

　　欠佳译文：在数控和自动化机床之间，有一个性能重复的广阔地带。

　　引申译文：数控和自动化机床有很多相同的性能。

(4) High-speed grinding does not know this disadvantage.

欠佳译文：高速磨床不知道这个缺点。

引申译文：高速磨床不存在这个缺点。

(5) You found that, in two experiments, hardness and greenness in apple went together with sourness.

欠佳译文：在有过两次经验之后，你发现苹果中的硬与青是和酸同行的。

引申译文：在有过两次经验之后，你发现凡是又硬又青的苹果总是酸的。

四、词性的变化

有些句子由于英汉两种语言的表达方式不同，故不能逐词对译。原文中有些词在译文中需要转换词性，才能使译文通顺自然。词性转移大体有以下几种情况：

1. 将英语动词、形容词、副词译成汉语名词

(1) Gases differ from solids in that the former has greater compressibility than the latter.

气体和固体的区别在于气体的可压缩性比固体大。

(2) The instrument is characterized by its compactness and portability.

这个仪器的特点是结构紧凑、携带方便。

(3) The cutting tools must be strong, tough, hard and wear resistant.

刀具必须有足够的强度、韧性、硬度，而且耐磨。

(4) Dynamics is divided into statics and kinetics, the former treating of force in equilibrium, the latter of the relation of force to motion.

力学分为静力学和动力学：前者研究平衡力，后者研究力和运动的关系。

(5) The image must be dimensionally correct.

图形的尺寸必须正确。

2. 将英语名词、介词、形容词、副词译成汉语动词

(1) The application of electronic computers makes for a tremendous rise in labor productivity.

使用电子计算机可以大大提高劳动生产率。

(2) In any machine input work equals output work plus work done against friction.

任何机器的输入功，都等于输出功加上克服摩擦力所做的功。

(3) Scientists are confident that all matter is indestructible.

科学家们深信一切物质是不灭的。

(4) Open the vales to let air in.

打开阀门，让空气进入。

3. 将英语名词、副词、动词译成汉语形容词

(1) The maiden voyage of the newly-built steamship was a success.

那艘新造轮船的处女航是成功的。

(2) It is a fact that no structural material is perfectly elastic.
事实上没有一种结构材料是十全十美的弹性体。

(3) They said that such knowledge is needed before they develop a successful early warning system for earthquakes.
他们说,这种知识对他们要发明一种有效的地震早期预报是必要的。

4. 将英语形容词、名词译成汉语副词

(1) A continuous increase in the temperature of the gas confined in a container will lead to a continuous increase in the internal pressure within the gas.
不断提高密封容器内气体的温度,会使气体的内压力不断增大。

(2) It is our great pleasure to note that China has joined the WTO.
我们很高兴地注意到,中国加入了世界贸易组织。

五、词义的差异

英语单词绝大多数为多义词,因此翻译时首先要选择一个最确切的词义。只有这样,才能使译文正确。选择词义通常从以下几个方面考虑:

1. 通过题材决定词义

题材是决定词义的第一步,相同的词用来谈论不同的话题,意思是不同的。以"Tension is building up."这句话为例,不同的题材下以下几个翻译都是正确的。

(1)(在一般生活中)形势紧张起来。
(2)(电学中)电压在增强。
(3)(力学中)张力在增大。
(4)(机械结构中)压力在增强。

2. 通过句法决定词义

句法指句子的排列组合规律,根据词在句中充当的不同成分来确定多义词的含义就相对容易多了。

例如,"like"在下面几个句子的翻译中词性不同,词义也就不同。

(1) Like charges repel, while unlike charges attract.
同性电荷相斥,异性电荷相吸。(这里"like"是形容词,作"同样的"讲)

(2) Things like air, water or metals are matter.
像空气、水或金属之类的东西都是物质。(这里"like"是介词,作"像"讲)

(3) Like knows like.
英雄识英雄。(这里"like"是名词,作"同类的人"讲)

(4) I hope I can use the computer like you do.
我希望能像你一样使用计算机。(这里"like"是连接词,作"如同"讲)

(5) Do you like this TV set?
你喜欢这台电视吗?(这里"like"是动词,作"喜欢"讲)

同一个词在不同的句子里,虽然词性相同,词义却可能有差别。在下面几个句子中,"light"均为形容词,但词义不同。

(1) The instrument is light.

这台仪器很轻。（这里"light"作"轻的"讲）

(2) The cover of the meter is light blue.

这个仪表盖是浅蓝色的。（这里"light"作"浅的"讲）

(3) The lamp is very light.

这盏灯很明亮。（这里"light"作"明亮的"讲）

3. 通过词义搭配决定词义

词与词在句子内部可以进行不同的搭配使用。当我们确定了其中一个词的意思后，就可据此来确定与其搭配的另一个词的意思。

例如，在不同的词组中，"fine"的意思不同。

(1) fine turning 微调（这里"fine"表示"turning"的范围）

(2) fine vacuum 高真空（这里"fine"表示"vacuum"的程度）

(3) fine copper 纯铜（这里"fine"表示"copper"的纯度）

4. 根据名词单、复数的不同选择词义

英语中有些词的单数和复数所表示的词义完全不同。例如：

名词	单数含义	复数含义
facility	简易,灵巧	设施,工具
charge	负荷,电荷	费用
main	主线,干线	电源
spirit	精神	酒精
work	功,工作	著作,工厂,工程

六、词的搭配

英语和汉语的词汇搭配规律不同，在翻译时需要考虑具体使用场合，根据上下文逻辑关系和各自语言的表达习惯，适当调整字面含义，进行搭配选择，而不能机械复制原文中的词汇搭配。

1. 动词与名词的搭配

以动词"move"为例，其基本含义是"运动"，但在不同的句中与不同的名词搭配时，其译法不同。

(1) The earth moves round the sun.

地球绕太阳旋转。

(2) Heat moves from a hotter to colder body.

热量从温度较高的物体传到温度较低的物体。

(3) Move M to other work.

调 M 去做别的工作。

(4) The work moves slowly.

工作进展缓慢。

2. 副词与动词的搭配

以副词"successfully"为例。

(1) Spot-welding has already been used successfully in welding fuel tanks.

  电焊已经成功地用来焊接油箱。

(2) The task was finished successfully.

  圆满完成了任务。

(3) Our production plan has been successfully carried out.

  我们的生产计划已顺利执行了。

3. 形容词与名词的搭配

以形容词"thick"为例。

(1) The book is 8 inches thick.

  这本书有8英寸厚。

(2) He drew a thick line on the paper.

  他在纸上画了一条粗线。

(3) Thick liquids pour much slowly than thin liquids.

  稠的液体比稀的液体流动性差。

(4) The conditions are too much thick.

  条件太过分了。

4. 动词与介词、副词的搭配

专业英语词汇中有大量动词与介词或副词搭配,这些搭配常为固定搭配,例如:set up(建立),differ from(与……不同),be characterized by(以……为特征),find out/work out(计算出,解决),be applied to(应用到),be divided into/be separated into(分成,分为),use up(用完,耗尽),give off(放出,长出),remove from(从……取下,脱下),study on/research on(研究),be based on(以……为基础),link to/connect with(连接),provide with(提供),be equipped with(装备,安装),go down/be decreased with/be reduced with(下降,减少),be increased with(增加),change into/turn into/convert into(变为),aim at(以……为目标),sum up(总结),lead to/result in(导致)。

附录2 科技英语翻译的标准与方法

一、翻译标准

  翻译的任务在于用一种语言文字忠实而流畅地传达另一种语言文字所包含的思想、内容。翻译包括两个阶段:理解和表达。透彻了解原文内容,把握其实质,再根据文章的类型特点采用不同的处理方法,保证内容和形式的有机统一。

  科技英语翻译是一门科学,不是艺术,不需要大量的再创作,译者只需准确、客观地反映所译学科的专业知识。在科技英语翻译中,准确是第一要素,如果为追求译文的流畅而牺牲准确,就会造成信息的丢失;另一方面,译文语言必须符合规范,遣词造句应符合本民族语言习惯,力求通俗易懂,不应有文理不通、逐字死译和生硬晦涩等现象。

科技文献主要为叙事说理,其特点一般是逻辑性强,结构严密,公式、数据和专业术语繁多,因此,科技英语翻译的标准应特别强调明确、通顺、简练。所谓明确,就是要求在技术内容上准确无误地表达原文的含义,做到概念清楚,逻辑正确,公式、数据准确无误,符合本专业要求,不应有模糊不清、模棱两可之处;所谓通顺,指遣词造句、译文语气应该恰当,尤其要恰当地表达出原文的语气、情态、语态、时态及所强调的重点;所谓简练,就是要求译文尽可能简短、精练,没有冗词、废词,在明确、通顺的基础上力求简洁明快、精练流畅。

因此科技英语的翻译应体现三个特点:专业性、客观性、精确性。

二、翻译方法与技巧

科技英语中词汇的翻译方法主要为直译(literal translation)和意译(free translation)。所谓直译,就是把原文中某个单词所指对象的名称换成译文的名称,也就是我们常说的词对词的翻译。所谓意译,就是将原文中出现的词汇的内涵与其所指对象综合考虑,在译文中用一个较恰当的词来替换原文中出现的词。

在科技英语的翻译中大量使用的是直译的方法,同时,意译也是翻译的重要手段,要求摆脱原文表层结构的束缚,根据句子的深层结构,对句子进行扩展、缩合、增词、省略、转换等处理,具体技巧论述如下:

(一)词的增译

词的增译就是为使译文忠实、通顺地表达原文意思,在译文中增加一些原文中无其形而有其义的词。增译是经常采用的翻译技巧之一。

1. 用汉语动词补充英语名词、动名词或介词的意义,使译文通顺

(1) The world needn't be afraid of a possible shortage of coal, oil, natural gas or other sources of fuel for the future.

世界无须担心将来可能出现煤、石油、天然气或其他燃料来源短缺的问题。

(2) Such a feat requires a perfect adjustment and timing of all the apparatus employed.

要做到这一点,需要所用设备的控制和定时准确无误。

(3) The molecules get closer and closer with the pressure.

随着压力增加,分子越来越接近。

2. 在表达动作意义的英语名词后增添汉语名词

(1) This lack of resistance in very cold metals may become useful in electronic computer.

这种在超低温中的金属没有电阻,可能对电子计算机很有用。

(2) Other flights going past Venus, and also Mars, are planned.

经过金星的,还有经过火星的其他飞行方案也计划出来了。

3. 增加表示名词复数的词

(1) The moving parts of a machine are often oiled so that friction may be greatly reduced.

机器的各个可动部件被经常加注润滑油,以便大大减小摩擦。

(2) The first electronic computers used vacuum tubes and other components and this made equipment very large and bulky.
第一代电子计算机使用电子管和其他元件,这使得设备又大又笨。

(3) Ground radar sets can "see" enemy planes at a great distance through darkness and clouds, and can direct antiaircraft to fire at them.
地面雷达装置可以透过夜幕或云层"看到"遥远的敌机群,并指示高射兵器向它们开火。

4. 增加某些被动语态或动名词中没有具体指出的动作执行者或暗含的逻辑主语

(1) The material is said to behave elastically.
据说,这种材料有弹性。

(2) To explore the moon's surface, rockets were launched again and again.
为了勘探月球的表面,人们一次又一次地发射火箭。

5. 在形容词前加名词

(1) According to Newton's Third Law of Motion action and reaction are equal and opposite.
根据牛顿运动第三定律,作用力和反作用力是大小相等、方向相反的。

(2) The washing machine of this type is indeed cheap and fine.
这种类型的洗衣机真是物美价廉。

6. 增加表示数量意义的概括性的词,起润色作用

(1) The frequency, the wavelength and the speed of sound are closely related.
声音的频率、波长与速度三者密切相关。

(2) A designer must have a good foundation in statics, kinematics, dynamics and strength of materials.
一个设计人员必须在静力学、运动学、动力学和材料力学这四个方面有很好的基础。

7. 增加使译文语气连贯的词

(1) Manganese is a hard, brittle, and gray-white metal.
锰是一种灰白的、又硬又脆的金属。

(2) In general, all the metals are good conductors, with silver the best and copper the second.
一般来说,金属都是良导体,其中银最佳,铜次之。

(3) Reading makes a full man; conference a ready man; and writing an exact man.
阅读使人充实;会谈使人敏捷;写作使人严谨。

(二)词的转译

英语和汉语属于不同的语系,英语属印欧语系,汉语属汉藏语系。不同语系的语言,无论在词汇还是语法方面都有很大的不同。就词性来说,同一意思在不同语言中可用不同词性来表达。翻译时适当改变原文中某些词的词性,以适应汉语的表达习惯,或达到一

定的修饰目的,这种翻译方法就叫词性转换。

1. 译成汉语动词

根据汉语在使用中的灵活性和广泛性的特点,除动词非谓语形式外,还可以把名词、形容词、介词译成汉语动词。

(1)将名词译成动词

①Were there no friction, transmission of motion would be impossible.

没有摩擦就不可能传递运动。

②The flow of electrons is from the negative zinc plate to the positive copper plate.

电子是从负的锌极流向正的铜极。

(2)将形容词译成动词

①When metal is cut, the shining surface is visible, but it turns gray almost immediately.

切削金属时,可以看到光亮的表面,但立刻就变成灰色。

②Nylon is nearly twice as strong as natural silk.

尼龙的强度约为真丝的两倍。

(3)将介词译成动词

①In general, positive or negative rake tool can be used on stainless steel.

通常正前角和负前角的刀具都可以用来加工不锈钢。

②Atomic power for ocean-going vessels is already a reality.

原子能动力用于远洋船只已成为现实。

2. 译成汉语名词

(1)将动词译成名词

①A voltmeter connected across A, B would read 10 volt.

接在 A、B 两点之间的伏特表的读数应当是 10 伏。

②Momentum is defined as the product of velocity and a quantity called the mass of the body.

动量的定义是物体的速度和质量的乘积。

(2)将形容词译成名词

科技英语中往往习惯于用表示特征的形容词及其比较级来说明物质的特性,因此在翻译时,可以在这类形容词后加"度""性"等词,使之成为名词。

①The more carbon the steel contains, the harder and stronger it is.

钢的含碳量越高,其强度和硬度越大。

②As most metals are malleable and ductile, they can be beaten into plates and drawn into wire.

由于大多数金属具有韧性和延展性,所以它们可以被压成薄板或拉成细丝。

3. 译成汉语形容词

(1)将副词译成形容词

①The electronic computer is chiefly characterized by its accurate and rapid computations.

电子计算机的主要特点是计算准确而且迅速。

②It is a fact that no structural material is perfectly elastic.

　　事实上,没有一种结构材料是十全十美的弹性体。

(2)将名词译成形容词

①The low stretches of the rivers show considerable variety.

　　河下游的情况是多种多样的。

②The electrical conductivity has great importance in selecting electrical materials.

　　在选择电气材料时导电性是很重要的。

(三)省略译法

　　省略译法也是英译汉经常采用的翻译技巧之一。英语中有些词如冠词、介词、连词、代词等,在汉语中没有或用得不多,这主要是因为汉语中可以借助于词序来表达逻辑关系。省略译法可以在使译文完全表达原文意思的前提下,把译文写得简明扼要、通顺流畅,更加符合汉语的表达习惯。

1. 省略代词

(1)By the word "alloy" we mean mixture of metals.

　　用"合金"这个词来表示金属的混合物。

(2)The volume of the sun is about 1,300,000 times that of the earth.

　　太阳的体积约为地球体积的130万倍。

2. 省略冠词

(1)The alternating current supplies the greatest part of the electric power for industry today.

　　如今交流电占了工业用电的绝大部分。

(2)Satellites can be sent into space with the help of rockets.

　　借助火箭可以把卫星送上太空。

3. 省略连词

(1)Give him an inch and he will take a mile.

　　他得寸进尺。

(2)If there were no heat-treatment, metals could not be made so hard.

　　没有热处理,金属就不会变得如此硬。

4. 省略介词

(1)They have found a method for solving this problem.

　　他们已经找到了解决这个问题的方法。

(2)Most substances expand in heating and contract in cooling.

　　大多数物质热胀冷缩。

5. 省略逻辑上或修辞上不需要的词

(1)As we know, electrons revolve about the nucleus, or center, of an atom.

　　正如我们所知,电子围绕着原子核旋转。

(2)A generator cannot produce energy. What it does is convert mechanical energy into electrical energy.

发电机不能产生能量,它只能把机械能转变为电能。

(四)反译法

反译法是指翻译时突破原文形式,如原文是肯定的,译文就用否定;如原文是否定的,译文就用肯定。这样做是为了达到一定的修辞目的。常用的反译法包括以下几种:

1. 将否定译为肯定

(1) Don't lose time in carrying out the plan.

立刻着手执行计划。

(2) There is no law that has no exception.

凡有规律都有例外。

(3) Metals do not melt until heated to a definite temperature.

金属要加热到一定的温度才熔化。

(4) In fact, there is hardly any sphere of life where electricity may not find useful application.

事实上,几乎任何一个生活领域中都要用到电。

2. 将肯定译为否定

(1) Worm gear drives are quiet.

蜗杆传动没有噪声。

(2) As rubber prevents electricity from passing through it, it is used as insulating material.

由于橡胶不导电,所以可作为绝缘材料。

(3) Early automobiles were sometimes only "horseless carriages" powered by gasoline or steam engines.

早期的汽车有时只不过是靠汽油或蒸汽机驱动的"无马马车"而已。

(4) The Theory of Relativity put forward by Einstein is now above many people's understanding.

对于爱因斯坦的相对论,现在还是有很多人不理解的。

3. 用双重否定来译一个肯定

一般是为了强调,但有时也可能是为了减弱或含有贬义。例如:

(1) Slowly he pulled the letter out of the envelope, and unfolded it.

他不慌不忙地从信封里抽出信纸打开来。

(2) We must put the factors into account.

我们不能不把这些因素考虑进去。

(3) The problem remains to be discussed in detail.

这个问题不能不详细讨论。

(五)被动语态的译法

专业英语中大量使用被动语态,这是因为科技文章需要描述客观事物的性质、特征、生产工艺流程与科学实验的结果。通常不需要说明从事这些活动的执行者。汉语中虽然也有被动语态,但适用范围比较窄。所以,在翻译英语的被动句时,应做一些改变,以适应汉语的表达习惯。

1. 译成汉语主动句

(1) Metal is widely used in industry.

　　金属广泛地应用在工业上。(无主语主动句)

(2) The mechanical energy can be changed back into electrical energy by means of a generator.

　　利用发电机,可以将机械能再转变成电能。(无主语主动句)

(3) All bodies are known to possess weight.

　　人们都知道所有物体都具有质量。(加施动者)

(4) The lathe is being adjusted by the operator.

　　操作员正在调整这台车床。(将主语译成宾语)

(5) The existence of organic chemistry is further justified by the importance of carbon compounds or mixtures in our lives.

　　有机化学的存在之所以必要,还因为碳的化合物或混合物在我们日常生活中非常重要。(无主语主动句)

2. 译成汉语被动句

用"被""由""受""靠""经""遭"等汉语中表达被动概念的介词引导出施动者。

(1) The tool is controlled by a code system.

　　刀具由代码系统控制。

(2) Heavier objects may be loaded with the assistance of other workers, autoloaders, a crane, or a forklift.

　　较重的工作可依靠其他工友、自动装卸料装置、吊车、铲车进行装载。

(3) Most wrought alloys are heat-treated and are not infrequently subjected to additional mechanical work after heat-treatment.

　　大部分锻造合金要经过热处理,并且在热处理之后还要经过后继的机械加工。

(4) Everything is built up of atoms.

　　万物都是由原子构成的。

3. 由 by 或 in 引导的状语可以转化为汉语的主语

(1) M codes are commonly used by the machine tool builder to give the user programmable ON/OFF switches for machine functions.

　　机床制造商通常用 M 代码给用户提供可编程的机床开/关功能。

(2) Storage capacity is generally provided in the MCU for any canned cycle.

　　通常机床数控装置(MCU)提供用于循环的存储空间。

4. 常用句型"It＋被动语态＋that"的译法

(1) It is believed that...　　　　人们认为……

(2) It is known that...　　　　众所周知……

(3) It is found that...　　　　人们发现……

(4) It is suggested that...　　　　人们建议……

(5) It is pointed out that... 有人指出……
(6) It is expected that... 人们希望……
(7) It is noted that... 人们注意到……
(8) It is announced that... 有人宣称……

(六) it 结构的译法

1. it 作代词

it 可以表示自然现象、天气、时间、距离等，是形式上的主语，没有词汇意义，翻译时可省略。it 还可以用来代替上文中提到的事物，翻译时可将其译为其所代替的事物。如：

The zener diode maintains the voltage across its terminals by varying the current that flows through it.

稳压管通过改变它的电流来维持其两端的电压。

2. it 作形式主语

it 作形式主语时，可代替不定式短语、动名词或主语从句。

(1) 代替不定式短语的句型

①it is＋形容词＋不定式

该句型最常见的变形形式是"it is＋形容词＋for...＋不定式"，不定式作主语时习惯放在句末，而句首用形式主语 it。这里的 for 引出不定式的逻辑主语，与不定式一起构成不定式复合结构，作主语放在句末。it 是引导词，作形式主语，其谓语动词是"系动词 be＋形容词表语"。这一结构相当于 that 引导的主语从句。翻译时一般有两种方法：一是把不定式复合结构译成汉语的主谓结构，仍作主语，即"……是 ……的"；二是把不定式的逻辑主语译作主语，而把形容词转成动词，与不定式一起译成谓语。如：

● It is important for a scientist to look at matter from the viewpoint of movement.
  科学家从运动的观点来看待问题是重要的。
● It is necessary for us to know how to convert energy.
  我们必须弄清楚能量是怎样转换的。

②it＋谓语动词＋不定式

该句型最常见的变形形式是"it＋谓语动词＋for...＋不定式"，这一结构与前一结构类似，都是不定式复合结构作主语放在句末，句首用引导词 it 作形式主语。这一结构的真正主语可以视情况翻译为"某人"或"某物"，也可以用不带逻辑主语的不定式，翻译时首先翻译主语，it 省略不译。如：

● It takes about 28 days for the moon to revolve around the earth.
  月亮绕地球一周约需 28 天。
● It required millions of years for fossil fuel to be formed from plants.
  从植物变成矿物燃料需要数百万年。

(2) 代替动名词的句型

动名词作主语，而用 it 作形式主语的结构不是很普遍，翻译时先译动名词。该句型最常见的形式是"it is no use＋动名词"，译为"……是无用的"。如：

①It is no use employing radar to detect objects in water.
使用雷达探测水下目标是不行的。
②It is no use learning a theory without practice.
学习理论而不实践是无用的。

(3)代替主语从句的句型

it 也是英语的多用词之一,它可以构成多种用法不同的结构。it 在这些结构中只有语法意义而无词汇意义,所以均省略不译,要是将其翻译成代词"它"或"这"就错了。

①it is(was)＋形容词＋主语从句。如：

It is certain that...　　　　……是确定的
It is desirable that...　　　 ……是理想的

②it＋不及物动词＋主语从句。如：

It seems that...　　　　好像是……
It turns out that...　　　显然……
It now appears that...　　现在看来,……
It follows that...　　　　由此可见/可以推断……

It follows that the greater conductance a substance has, the less is the resistance.
由此得出,一种物质的导电率越大,电阻越小。

③it＋谓语动词的被动语态＋主语从句。如：

It is reported that...　　　　　　据报道,……
It is generally recognized that...　大家公认为……
It is supposed that..　　　　　　假设……
It is assumed that...　　　　　　假定……
It is found that...　　　　　　　 人们发现……

It has been found that a force is needed to change the motion of a body.
人们发现,要改变一个物体的运动状态,就需要加外力。

④it is＋名词＋主语从句。如：

It is a pity that...　　　　　　　　遗憾的是……
It is common knowledge that...　常识是……

3. it 作形式宾语

当宾语较长且带有补语时,通常 it 作形式宾语,代替不定式、动名词或宾语从句,放在谓语动词之后,而将真正的宾语放在补语之后,以保持句子的平衡。使用这一结构的动词有 think、make、find、consider、believe 等。常见句型如下：

(1)动词＋it＋形容词＋that

翻译此句型时,形式宾语 it 不译,把宾语从句译成汉语的兼语式,即宾语兼主语,或者把从句与形容词一起译成宾语。如：

①Do you think it possible that the electronic computer will replace man?
你认为电子计算机取代人是可能的吗?

②People consider it true that all types of radiant energy move through a vacuum at the speed of light.

人们认为各种辐射能确实是以光速通过真空管的。

(2)动词＋it＋形容词＋不定式

该形式也可变形为"动词＋it＋形容词＋for＋不定式"，for 引导不定式的逻辑主语，真正的宾语是不定式复合结构。如：

①Heat from the sun makes it possible for plants to grow.

来自太阳的热量使植物的生长成为可能。

②When we want to measure very small currents we find it convenient to use milliampere and microampere.

当要测量很小的电流时，我们觉得用毫安和微安是比较方便的。

(3)动词＋it＋形容词＋动名词

该形式不常见，其翻译方法基本相同。如：

①We find it useless employing radar to detect objects in water.

我们发现使用雷达探测水下目标是不行的。

②People found it possible exerting a large force by the application of a small force.

人们发现有可能用一个小的力来产生一个大的力。

(4)动词＋介词＋it＋that

该形式中 it 也是一个引导词，但其作用是充当介词的形式宾语，真正的宾语仍是 that 从句。当 that 从句作某些动词的宾语时，一般不直接跟在介词之后，而用一个形式宾语 it 插在介词和 that 之间。翻译时，it 省略不译，that 从句仍译成宾语。如：

①We will answer for it that this test is reliable.

我们保证这项试验是可靠的。

②This is why every aircrew member must see to it that this shape is maintained as accurately as possible.

这就是为什么每个飞行员都必须使(机翼的)这种形状尽可能保持正确的原因。

4.it 用于强调句型

强调句型是简单句，可用于强调句中的主语、宾语和状语，但不能强调谓语和定语，强调句的句型为：It is(was)＋被强调的成分＋that(who)。在这种句型中，it 和 that 都没有词汇意义，翻译时可在强调成分前加上"正是""就是"等。如：

(1)It is in the form of alloys that metals are often used in industry.

在工业中就是经常以合金的形式使用金属。

(2)It was not until 1886 that aluminum came into wide use.

直到1886年，铝才得以广泛的应用。

(七)句子成分转译

英语和汉语由于表达方式不同，翻译时往往需要改变原文的语法结构。其所使用的主要方法除了词类转换之外，还有句子成分的转换。在一定情况下，适当改变原文的某些句子成分,可达到译文逻辑正确、通顺流畅、重点突出等目的。

1. 将介词宾语译成主语

英译汉时,为了符合汉语的表达方式,有时需要将原文中的介词宾语翻译为译文中的主语,使译文重点突出、行文流畅。如:

(1) Iron comes between manganese and cobalt in atomic weight.

铁的原子量在锰和钴之间。

(2) The electric arc may grow to an inch in length.

电弧长度可以增长到1英寸。

(3) A motor is similar to a generator in construction.

电动机的结构与发电机类似。

(4) Our refrigerator is light in weight, portable in size, and low in power consumption.

我们生产的电冰箱质量轻、体积小、耗电低。

2. 将动词 have 的宾语译成主语

(1) The proton has considerably more mass than the electron.

质子的质量比电子大得多。

(2) Levers have little friction to overcome.

杠杆要克服的摩擦力很小。

(3) Evidently semiconductors have a lesser conducting capacity than metals.

半导体的导电能力显然比金属差。

(4) Water has a density of 62.5 pounds per cubic foot.

水的密度是每立方英尺62.5磅。

3. 将其他动词的宾语译成主语

除动词 have 的宾语外,其他及物动词的宾语有时也可译成主语,译法基本相同,不过动词(省略不译除外)有时需要与主语一起译成定语。如:

(1) Hot-set systems produce higher strength and age better than cold-set systems.

热固系统比冷固系统的强度高,而且老化情况也比较好。

(2) We need frequencies even higher than those we call very high frequency.

我们所需的频率甚至比我们称之为超高频的频率还要高。

(3) Light beams can carry more information than radio signals.

光束运载的信息比无线电信号运载的信息多。

(4) Most cylinder cushions provide a relatively small velocity change.

大多数缸体缓冲器所引起的速率变化均较小。

4. 将主语译成定语

翻译时,往往由于更换主语,而将原来的主语译成定语。

(1)将形容词译成名词

将形容词译成名词并译作主语时,原来的主语通常都需要译成定语。如:

①Medium carbon steel is much stronger than low carbon steel.

中碳钢的强度比低碳钢大得多。

②The wings are responsible for keeping the airplane in the air.

机翼的用途是使飞机在空中保持不下坠。

③The oxygen atom is nearly 16 times as heavy as the hydrogen atom.

氧原子的质量几乎是氢原子的16倍。

④This steam engine is only about 15 percent efficient.

这种蒸汽机的效率只有15％左右。

(2)将动词译成名词

将动词译成名词并译作主语时,原来的主语一般需要译成定语。如:

①The earth acts like a big magnet.

地球的作用像一块大磁铁。

②Mercury weighs about 13 times as much as water.

水银的质量约为水的13倍。

③The vertical spindle-drilling machine is characterized by a single vertical spindle rotating fixed position.

立式钻床的特点是具有一根单独在固定位置上旋转的垂直主轴。

④A cathode ray tube is shaped like a large bell.

阴极射线管的形状像个大铃铛。

5. 将定语译成谓语

(1)将动词宾语的定语译成谓语

①Copper and tin have a low ability to combine with oxygen.

铜和锡的氧化能力低。

②Water has a greater heat capacity than sand.

水的热容比沙大。

③Neutron has a mass slightly larger than that of proton.

中子的质量略大于质子的质量。

④A semiconductor has a poor conductivity at room temperature, but it may become a good conductor at high temperature.

在室温下,半导体的导电率低,但在高温下,它可能成为良导体。

(2)将介词宾语的定语译成谓语

在将介词宾语译成主语的同时,有时还需要把该宾语的定语译成谓语,原来的谓语都译成定语。如:

①Gear pumps operate on the very simple principle.

齿轮泵的工作原理很简单。

②Nylon is produced by much the same process as rayon.

尼龙的生产过程与人造丝大体相同。

③Though each cam appears to be quite different from the others, all the cams work in a similar way.

虽然每种凸轮都不大相同,但所有凸轮的运作方式都相同。

④Radar works in very much the same way as the flashlight.

雷达的工作原理和手电筒极为相似。

6. 译成汉语主谓结构中的谓语

有时出于修辞的目的,将某一名词前面的形容词(即名词的定语)与该名词颠倒顺序,一起译成汉语的主谓结构,在句子中充当一个成分,原来作定语的形容词成为主谓结构中名词的谓语。如:

(1) Among the advantages of numerical control are more flexibility, higher accuracy, quicker changes, and less machine down time.

数控的优点是适应性好、精度高、变化快、非加工时间短。

(2) These pumps are featured by their simple operation, easy maintenance, low oil consumption and durable service.

这些水泵的特点是操作简单、维修容易、耗油量少、经久耐用。

(3) Other requirements of the lathe tool are long life, low power consumption, and low cost.

车刀的其他要求是使用寿命长、动力消耗少并且造价低。

(4) Briefly, a long thin wire has a high resistance, and short thick wire has a low resistance.

简单地说,长而细的导线电阻大,短而粗的导线电阻小。

(八)虚拟语气的翻译

虚拟语气是说话人想表达主观愿望和假想虚拟的情况或表示一种委婉的口气时,英语中所使用的一种特殊的谓语动词形式。在科技英语中也常有这种情况,翻译时注意要将其与陈述语气加以区分。

1. 虚拟条件从句

虚拟条件从句是指假设的情况根本不可能存在,或者发生的可能性很小,这时采用虚拟语气。它的基本形式有如下两种:

(1)表示与现在的事实相反的假设

从句:谓语动词用过去时,be 一律用 were。

主句:should(would, could, might)+动词原形。

如:If no force acted on the moon it would move in a straight line with constant speed.

如果月球不受力的作用,它就会以恒速做直线运动。(这里的条件根本不可能存在)

(2)表示与过去的事实相反的假设

从句:had+过去分词。

主句:should(would,could,might)+have+动词过去分词。

如:If there had been no radio or television, man could not have seen the far side of the moon.

如果没有无线电和电视,人们就不能看到月球的背面了。

(3)省略 if 的倒装虚拟条件句

在 if 引导的条件状语从句中有 were、had 时,有时可省略 if 而采用倒装语序,此时可将 were、had 提到主语之前。如:

Had radio not been invented, many jobs, such as helping planes to land in bad weather and calling doctors to far-off places, would not have been possible.

如果没有发明无线电,很多工作,如在恶劣的天气下帮助飞机着陆、请医生到遥远的地方去等就做不成了。

(4)引导虚拟条件从句的连词

通常将 if 作为虚拟条件从句的连词。此外,suppose、provided(that)、once、unless 等连词都可引导虚拟条件从句。如:

From Lenz's Law we knew that neither electrical nor other forms of energy can be derived unless an equivalent amount of work were performed.

根据楞次定律我们知道,如果不做等量的功,就既不能得到电能,又不能得到其他形式的能量。

2. 谓语动词一律用原形的虚拟条件句

在表示现在或将来有可能实现的一种假想时,从句中一切人称后的谓语都用动词原形,而主句中的谓语则用陈述语气,这种用法在科技语中很常见。如:

It follows that those points must be at the same electrical level if there be no movement of electricity between any two points.

如果任意两点间没有电荷流动,则这两点必须处于相同的电位。

3. 虚拟语气用于一些从句的情况

(1)在表示愿望、建议、要求等动词后的宾语从句中常用虚拟语气,如 wish、desire、require、demand、think、expect、suggest、propose、imagine 等,具有愿望、建议、命令的含义,它们后面的从句中谓语动词常用"(should)+动词原形",且一般都省略 should。如:

He suggested that we operate computers this afternoon.

他建议我们今天下午操作计算机。

(2)as if、as though(好像,仿佛)引导的方式状语从句用虚拟语气。如:

You can be heard just as clearly as if the two of you were in the same room.

声音听起来很清楚,就好像你们两个人在同一个房间里一样。

4. 虚拟假设情况的其他表示方法

虚拟条件句的假设情况还可用下列含有条件意味的词、短语或句子来表示:or less(否则,要不然),without(如果没有),otherwise(否则),in the absence of(如果不存在,如果没有),if it were not for(if there were no...)等。如:

If it were not for friction, we could not walk and automobiles could not run.

如果没有摩擦力，我们就不能走路，汽车就不能行驶。

(九)非谓语动词的翻译

非谓语动词结构在科技英语中应用广泛，它是为了保证科技文体的行文简洁、结构紧凑而将分句缩略而成的。非谓语动词是动词的一种非限定形式，不可单独作谓语，其形式不受主语的人称和数的限制，有 V-ing、V-ed、to-V 三种形式。由于非谓语动词由动词变化而来，因此保留了动词的特性，可带有自己的宾语和状语。

1. 非谓语动词 V-ing 的翻译

V-ing 形式通常被划分为动名词和现在分词，使用时也可以不加区分，但要准确地把握其用法，弄清楚 V-ing 形式在句中的作用及意义。V-ing 形式可以在句中作主语、表语、宾语、定语和状语等。

(1)作主语

①Knowing how to use computer is useful.

知道怎样使用计算机是很有用的。

②Controlling a machine tool by means of a prepared program is known as numerical control(NC).

用准备好的程序控制机床，这种工作方式被称为数字控制(NC)。

(2)作表语

①His great fun is chatting with strangers online.

他最大的乐趣就是在网上和陌生人聊天。

②The present situation is encouraging.

当前的形势令人鼓舞。

(3)作宾语

①It may also have a polarity switch to facilitate reversing the test leads.

还有一个极性开关可以用来很方便地交换测试笔的极性。(作"to facilitate"的宾语)

②The second step in minimizing costs of production is choosing the cheapest of the technical efficient alternatives.

降低生产成本的第二个措施是选用技术上有效的最便宜的替代品。(作"in"的宾语)

(4)作定语

单个的 V-ing 作定语时，常放在被修饰的名词的前面(也可放其后)，但 V-ing 短语作定语一般放在被修饰的名词之后。如：

①All moving bodies have energy.

所有运动的物体都有能量。

②A direct current is a current flowing always in the same direction.

直流电流是指总向同一个方向流动的电流。

(5)作状语

V-ing 形式作状语时，常表示主句谓语动词的动作发生的时间、原因、条件、结果、方

式、让步、伴随情况等,它可以置于句首、句中或句尾。如:

①Lifting something, you do work.

当你举起物体时,你就在做功。(条件状语)

②When measuring current, the circuit must be opened and the meter inserted in series with the circuit or component to be measured.

测量电流时必须断开电路,将万用表与待测电路或元器件串联。(时间状语)

(6)作主语或宾语的补足语

We put a hand above an electric fire and feel the hot air rising.

我们把手放在电炉的上方,就会感觉到热空气在上升。

(7)V-ing 的独立主格结构

一般情况下主句的主语就是 V-ing 形式的逻辑主语,但有时 V-ing 形式的逻辑主语和主句主语并不一致,V-ing 形式可以有其独立的逻辑主语。如:

①Weather permitting, we'll have the match tomorrow.

天气允许的话,明天我们将举行比赛。(条件状语)

②With Mr. Smith taking the lead, they decided to set up a trading company.

以史密斯先生为首,他们决定成立一个贸易公司。(方式状语)

2. 非谓语动词 V-ed 的翻译

非谓语动词 V-ed 形式含有被动与完成之意,在句中主要起形容词和副词的作用,可在句中担任定语、表语、状语和补语等。

(1)V-ed 作定语

单个 V-ed 作定语时,其位置比较灵活,常放在被修饰名词的前面,也可放在被修饰名词的后面。V-ed 短语作定语一般放在名词之后,相当于一个定语从句。如:

Heat is the energy produced by the movement of molecules.

热是分子运动所产生的能量。

(2)V-ed 作表语

如:The students will get confused if they are made to learn too much.

如果让学生学得太多,他们会感到糊涂的。

(3)V-ed 作状语

V-ed 作状语表示时间、原因、条件、方式、让步等。如:

①Observed from the spaceship, our earth looks like a blue ball.

从宇宙飞船上看,我们的地球像一个蓝色的球体。

②Tonight, lit by countless electric lights, all the halls are as bright as day.

今晚无数个电灯把整个大厅照耀得如同白昼。

3. 非谓语动词 to-V 的翻译

非谓语动词 to-V 在句中可以起名词、形容词及副词的作用,可以用作主语、宾语、表语、定语和状语等,常用来表示具体的(特别是未来的)一次性动作。如:

(1) It is easier to perform quality assurance by a spot-check instead of checking all parts.

通过抽检代替全检更容易保证质量。(作主语)

(2) To describe the motion one must introduce the concept of time.

为了描述运动,必须引入时间的概念。(作状语)

(十)否定形式的翻译

英语的否定形式是一个常见而又比较复杂的问题,使用非常灵活、微妙,被认为是英语的特点之一。否定形式在科技英语中的应用很广泛。在表达否定概念时,英语在用词、语法和逻辑等方面与汉语有很大的不同。有的句子形式上是肯定的而事实上是否定的,而有的句子形式上是否定的但事实上却是肯定的。英语否定词的否定范围和重点有时难以判断,否定词在句中表示强调的方法与汉语不同,某些否定词和词组的习惯用法较难掌握。因此,在翻译否定形式时,必须细心揣摩,真正彻底理解其意义及否定的重点,然后根据汉语的习惯进行翻译。如:

(1) The earth does not move round in the empty space.

(误)地球在空无一物的空间中不运转。

(正)地球不是在空无一物的空间中运转。

(2) All metals are not good conductors.

(误)所有金属都不是良导体。

(正)并非所有金属都是良导体。

从上面例句中可以看出,英语的否定形式与否定概念并非永远一致。例(1)中,形式上是一般否定(谓语否定),但实际上却是特指否定(其他成分否定);例(2)中,句子看似全部否定,但却是部分否定。

1. 否定成分的转译

一般说来,否定形式仍翻译成否定形式。但由于英语和汉语两种语言表达方式和习惯的不同,有些否定形式应译成肯定形式。这种正反、反正表达法是翻译的一种重要技巧。有些英语否定句虽然是用一般否定(否定谓语)的形式,但在意义上却是特指否定,即其他成分的否定;反之,有些句子形式上是特指否定,而意义上却是一般否定,翻译时要根据汉语的习惯进行否定成分的转译。如:

(1) Sound does not travel so fast as light.

声音不像光传播得那样快。(原文否定谓语,译文否定状语)

(2) Neutrons carry no charge.

中子不带电荷。(原文否定宾语,译文否定谓语)

(3) The sun's rays do not warm the water so much as they do the land.

太阳光线使水增温,不如它使陆地增温那样高。(原文否定谓语,译文否定状语)

(4) Matter must move, or no work is done.

物质必须运动,否则就没有功。(原文否定主语,译文否定谓语)

(5) The mountain is not valued because it is high.

山的价值并不是因为它高。(原文否定谓语,译文否定谓语)

(6) Green plants cannot grow strong and healthy without sunlight.

没有阳光,绿色植物就长不结实,长不好。(原文否定谓语,译文否定谓语)

(7) We do not consider melting or boiling to be chemical changes.

我们认为熔化或沸腾不是化学反应。(原文否定谓语,译文否定兼语式的第二谓语)

2. 部分否定的译法

英语的否定有全部否定与部分否定。全部否定指否定整个句子的全部意思,可用 none、neither、no、not、nothing、nobody 等否定词。部分否定主要由 all、every、both、always 等含全体意义的词与否定词 not 构成,其表达的意义是部分否定,相当于汉语"不是所有都""不是两者都""不总是"之意。当将否定词 not 放在这些词之前(not all, not every, not both)时,其部分否定的意义就很明显,一般不会翻译错。然而,否定词 not 有时却与谓语在一起,构成谓语否定,形式上很像全部否定,但实际上却是部分否定,翻译中应特别注意。如:

(1) All minerals do not come from mines. (=Not all minerals come from mines.)

并非所有矿物都来自矿山。(不是"所有矿物都不来自矿山。")

(2) Every color is not reflected back. (=Not every color is reflected back.)

并非每种色光都会反射回来。(不是"每种光都不反射回来。")

(3) Both of the substances do not dissolve in water.

不是两种物质都溶于水。

(4) But friction is not always useless. In certain cases it becomes a helpful necessity.

摩擦并非总是无用的,在某些场合下,它是有益的、必需的。

(5) All the chemical energy of fuel is not converted into heat.

并非所有燃料的化学能都转变成热量。

3. 否定语气的改变

英语的否定句并非一概译成汉语的否定句,因为英语中有些否定句表达的是肯定的意思,还有些否定句在特定场合下可以表达肯定的意思。例如 nothing 在下列句中的译法:

(1) Energy is nothing but the capacity to do work.

能就是做功的能力。

(2) An explosion is nothing more than a tremendously rapid burning.

爆炸仅仅是非常急速的燃烧。

(3) Ball bearings are precision-made bearings which make use of the principle that "nothing rolls like a ball".

滚珠轴承是精密轴承,采用了"球形最善于滚动"的原理。

4. 否定意义的表达

英语中还有许多肯定句,所表达的却是否定的意思。在这类句子中虽然没有出现否定词,但句子中有些词却含有否定的意义,翻译时一般都要将否定意义译出,译成汉语的否定句。对这类句子的理解和翻译,只要掌握了词汇意义,就不会有什么困难。常见的含有否定意义的词和词组有:little(几乎没有),few(几乎没有),seldom(极少),scarcely(几乎不)，hardly(很难)，barely(几乎没有)，rarely(很少)，too...to(太……不)，too...for(太……不)，free from(没有)，fall short of(没有达到)，fail to(不能)，instead of(而不是)，far from(完全不)，in the absence of(没有……时)。如：

(1) Of all metals silver is the best conductor, but it is too expensive to be used in industry.

所有金属中，银是最好的导体，但成本太高，不能在工业上使用。

(2) The distance from the sun to the earth is too great to imagine.

太阳到地球的距离大得不可想象。

(3) The angularity of the parts is too great for proper assembly.

零件的斜度太大，不适于装配。

(4) If the follow loses contact with the cam, it will fail to work.

随动元件如果与凸轮脱开，就不能工作。

5. 双重否定的译法

英语和汉语一样，也有双重否定结构。英语的双重否定是由两个否定词(no, not, never 等)连用或一个否定词与某些表示否定意义的词连用而构成的。双重否定表示否定之否定，即强调肯定，因此翻译时有两种译法，既可以译成双重否定，又可以译成肯定。

双重否定从语气的强弱上分，有弱化的双重否定和强化的双重否定两种。弱化的双重否定中包含的两个否定词，一个常由否定前、后缀构成。由于两个否定词中一个为另一个所否定，使否定的语气弱化，从而将否定的效果抵消了一部分。

(1) 译成双重否定

① There is no steel not containing carbon.

没有不含碳的钢。

② Sodium is never found uncombined in nature.

自然界中从未发现不处于化合状态的钠。

③ No flow of water occurs unless there is a difference in pressure.

没有压力差，水就不会流动。

④ It is impossible for heat to be converted into a certain energy without something lost.

热转换成某种能而没有什么损耗是不可能的。

(2) 译成肯定

① There is no law that has no exceptions.

凡是规律都有例外。

② A radar screen is not unlike a television screen.

雷达荧光屏跟电视荧光屏一样。

③ There is nothing unexpected about it.

一切都在意料之中。

④ One body never exerts a force upon another without the second reacting against the first.

一个物体对另一物体施加作用力就必然会受到另一物体的反作用力。

（十一）长难句的翻译

科技英语文体的特点是清晰、准确、精练、严密。在科技英语的句子中常带有许多修饰、限定成分。大量长句的使用是科技英语文章的特点之一。由于其内容、使用域和语篇功能的特殊性，其长句在许多方面有别于日常英语，其差别主要表现在句法和词汇上。英语长句主要长在修辞成分上。英语句子的修辞成分主要是名词后面的定语短语或定语从句，以及动词后面或句首的介词短语或状语从句。英语长句一句话可以表达好几个意思，而汉语习惯用一个小句表达一层意思，一般好几层意思要通过几个小句来表达。翻译时，有必要通过句法分析找出句子的主要成分，弄清主要成分与各修辞成分之间的逻辑关系，分清层次，再根据汉语的表达习惯重新组句。长难句的翻译方法主要有：

1. 次序的处理

(1) 顺译法

在科技英语中，有些长句叙述的内容基本上是按照所发生的时间先后安排，或者按照逻辑关系安排，与汉语的表达方法比较一致，因此可以按照原文的顺序译出。这种长句的拆译称为顺译法。如：

①Some of these causes are completely reasonable results of social needs. Others are reasonable consequences of particular advances in science being to some extent self-accelerating.

在这些原因中，有些完全是自然而然地来自社会需求，而另一些则是由于科学在一定程度上自我加速而产生某些特定发展的必然结果。

②However, fiber systems can carry so much more telephone conversations at the same time than wire pairs, and can carry them so much farther without amplification of regeneration, that when there are many telephone calls to be carried between points such as switching offices fiber systems are economically attractive.

然而，光纤系统同时可容纳比线对多许多的电话线路，并且在不需要放大的情况下把信号送得更远。当两点之间，比如中继局有很多电话呼叫要被传送时，光纤系统就是非常经济的。

(2) 倒译法

英语与汉语的表达习惯不同，英语习惯于前置性陈述，即先结果后原因，而汉语则一般先原因后结果。当英语中的逻辑次序和汉语相反时，例如表示条件、时间、让步等意义的从句位于主句之后时，往往将从句提到主句之前，从长句的后面或中间译起，把长句的开头放在译文的结尾，这种译法称为倒译法。如：

①Additional social stresses may also occur because of the population explosion or problems arising from mass migration movements—themselves made relatively easy nowadays by modern means of transport.

由于人口的猛增或大量人口的流动(现代交通工具使这种流动相对容易)所造成的种种问题也会对社会造成新的压力。

②This is why the hot water system in a furnace will operate without the use of a water pump, if the pipes are arranged so that the hottest water rises while the coldest water runs down again to the furnace.

如果把管子装成能让最热的水上升而让最冷的水流回到锅炉去的结构,那么锅炉中的热水系统不用水泵就能运转,其道理就在于此。

2. 结构的处理

(1)采用短句翻译

有些长难句或介词短语等修饰语与主句的关系不十分密切,可遵照汉语多用短句的习惯,将它们译成短分句分开叙述。如:

The laser, its creation being thought to be one of today's wonders, is nothing more than a light that differs from ordinary lights only in that it is many, many times more intense and so can be applied, in the fields that no ordinary light has ever been able to penetrate into.

激光的发明虽然被认为是当代的一个奇迹,但其实它就是一种光。这种光不同于普通的光是因为它比普通的光强烈许多倍,因而能够应用于普通光无法穿透的地方。

(2)简化句子翻译

翻译时有时可以将从句压缩成一个词或词组,使句子更为精练。如:

①With radar, we can "see" things at a great distance, and it shows us how far away they are, in which direction they lie, and what movements they are making.

有了雷达,我们就能"看到"远方的物体,雷达还能向我们指明这些物体的距离、方位和运动方式。

②For use where small size, light weight and portability are highly desirable, transistorized oscilloscopes are preferred over vacuum-tube type.

在应用中如果需要体积小、质量轻和携带方便,那么采用晶体管示波器比采用电子管优越。

3. 综合法

有些长难句结构复杂,单独使用前述的任何方法都不合适,这时需要仔细分析原文,按照汉语习惯进行综合处理。如:

(1)Noise can be unpleasant to live even several miles from an aerodrome; if you think what it must be like to share the deck of a ship with several squadrons of jet aircraft, you will realize a modern navy is a good place to study noise.

噪声甚至会使住在远离飞机场几英里以外的人感到不适。如果你能想象到站在甲板上的几个中队喷气式飞机中间将是什么滋味的话,那你就会意识到现代海军是研究噪声的理想场所。

(2) The super-cooling effects of the cryogenics which convert liquid helium and

other gases into "superfluids" and metals into "superconductors", making them non-resistant to electricity, could change the world in a number of ways.

低温学的过冷作用将液态氦及某些气体变成"超流体",将某些金属变成"超导体",使它们没有电阻,从而可以使用许多方法改变世界。

(十二)专业术语的翻译

科技术语是科技文献中的专业性词汇,这类术语与一般词语相比具有更精确、更固定的含义。随着科技的发展,新的发明创造不断涌现,新的科技术语不断出现,翻译科技术语时必须首先弄清原词的专业含义,再选择或转译为相应的汉语术语。

1. 译名的统一

科技术语一般只表示一个特定的概念,所以大多数科技术语是单义的。但是科技术语译名翻译时不统一、不规范,会给读者带来困扰,因此在翻译科技术语时,要尽量采用有关领域公布的统一译名,或者已为大家公认的译名,而不应采用不常见的或已过时的译名。如:lathe 一词应译为"车床",而不应译为"旋床";cement 曾译为"士敏土""水门汀"等,现应通译为"水泥"。再如:spindle nose 应译为"主轴头"或"主轴端",如将其译为"主轴鼻"就不易理解了;nylon 现在习惯译为"尼龙",就不能将其随便音译为"乃隆"。

如果某专业术语有众多译名,就应查阅相关的专业资料,确定相应的规范译名。如 saddle 一词可以译为"鞍座""床鞍""鞍板"等,那么在车床中应如何翻译呢?查阅金属切削方面的专业书籍,译为"溜板"为好。

还有一些科技术语有多种习惯译法,在此情况下,在同一本书或同一篇文章中应保持通用译名的统一性,对于该词的翻译应保持前后一致。如 nylon 一词可译为"尼龙""耐纶""酰胺纤维"等通用译名,若前面译为"尼龙",后面译为"耐纶"或"酰胺纤维",就可能使读者误认为是不同的材料。

2. 科技术语的译法

科技术语的翻译同样有直译和意译两种,其中直译采用较多,以便读者直接理解术语的确切含义。具体的直译方法有移植(transplant)、音译(transliteration)、形译(pictographic translation)等;具体的意译方法有推演(deduction)、引申(extension)等。有时采用综合译法,即部分直译、部分意译。其基本译法举例说明如下:

(1)移植

移植就是按词典所给出的解释将词的各个词素的含义依次翻译出来。这种方法多用于复合词和派生词。

①复合词,即由两个或更多的词组成的专业技术词汇。如:microwave(微波),de-oil(脱脂),linear-induction motor(线性马达),mechatronics(机电一体化)。

②派生词,即在已有词汇的单词或词干的前面或后面通过加词缀的方法来构成新词。如:micromachining(微切削加工),machinability(可加工性),superconductor(超导体),coaxial(同轴的)。

(2)音译

专有名词(如地名、人名、计量单位等)通常需要音译。除此之外,有些用来表示新材料、新产品、新概念、新理论的词,或是一些表示药名、商标名称、机械设备名称的词以及一些缩略名词,也可以借助于音译。

①计量单位一般用音译。如:ohm(欧姆),watt(瓦特),ampere(安培),volt(伏特)。

②新名称术语的音译。如:aspirin(阿司匹林),nylon(尼龙),sonar(sound navigation ranging)(声呐),microphone(麦克风)。

③音义结合。如:jeep(吉普车),valve body(阀体),Norton gear(诺顿齿轮),Taylor formula(泰勒公式)。

(3)形译

所谓形译指根据物体的形状对其进行解释翻译。

①选用与字母形状相似的汉语的词来译。如:T-beam(丁字梁),I-steel(工字钢),T-square(丁字尺),T-wrench(丁字扳手)。

②选用能表达原文字母或单词形象的汉语来译。如:U-steel(槽钢),V-belt(三角皮带),herringbone gear(人字齿轮),U-shaped magnet(马蹄形磁铁)。

③保留字母不译,在该字母后加"形"字。如:A-frame(A形架),C-clamp(C形夹),D-valve(D形阀),O-ring(O形环)。

④保留字母不译,字母代表某种含义。如:X-ray(X射线),P-region(P区,电子不足区),N-region(N区,电子剩余区)。

(4)推演

所谓推演,即根据词典的字面含义,结合原文的具体语境推断出词义。如 space shuttle(航天飞机),space 的字面含义为"太空"或"空间",shuttle 的字面含义为"织布用的梭子",若译为"太空穿梭机",则易引起误会,通译为"航天飞机",既直观又易懂。

(5)引申

所谓引申,是指在不脱离原文的基础上,运用延续或扩展的方法进行翻译。如:brain-trust(智囊团),将 brain 的具体含义"大脑"引申为"智力";data qualification(数据限定条件),将 qualification 的抽象含义"鉴定过程"缩小为"通过鉴定程序必须具备的具体条件"。

3. 机器设备常用说明文字及设备装箱标识用语的翻译举例

随着进出口设备的不断增多,机器设备及其装箱的标识常用英语表达,这些常用语的表达简洁明了,下面列出一些简单的例子以供参考。

手动、自动:MANUAL;AUTO

(电流)导通、断路:ON;OFF

入口、出口:INLET;OUTLET

纵向、横向:LONGITUDINAL;TRANSVERSE

(离合器)接、脱开:ENGAGE;DISENGAGE

启动、停车:START;STOP

开、关:OPEN;CLOSED(常用于一般机械装置)/ON;OFF(常用于电源开关)

注油:PRIME

加机油点:OIL
加黄油点:GREASE
勿受潮:KEEP DRY
勿倒置:KEEP UPRIGHT
切勿倾倒:DON'T TURN UPSIDE DOWN
切勿平放:NOT TO BE LAID FLAT
切勿挤压:DON'T CRUSH
勿放在重物之下:NOT TO BE STORED UNDER HEAVY CARGO
竖放:STAND ON END
怕光:KEEP IN DARK PLACE
怕热:PROTECT AGAINST HEAT
此处吊起:SLING HERE 或 LIFT HERE
此处打开:OPEN HERE
高压危险:HIGH TENSION
此盖打开前切断电源:MAKE SURE POWER IS OFF WHEN REMOVING THIS COVER
注意:如果仪器长期不用或电池已用完,应将电池取出。
CAUSION:REMOVE THE BATTERIES WHEN THE UNIT IS TO BE STORED FOR LONG PERIODS OF TIME OR IF THE BATTERIES BECOME EXHAUSTED.

附录3 机电专业基础词汇表

※ 基本电路 ※

conductor 导体
insulation 绝缘
insulator 绝缘体
source 电源
battery 电池
switch 开关
positive pole 正极
negative pole 负极
circuit 电路
open circuit 开路
short circuit 短路
closed circuit 闭合电路;通路
parallel connection 并联

series connection 串联
current 电流
Ohm's Law 欧姆定律
direct current(DC) 直流电
alternating current(AC) 交流电
voltage 电压
electromotive force(EMF) 电动势
terminal voltage 端电压
low-voltage 低压
high-voltage 高压
electric potential 电位
electric potential difference 电位差
potential 电势

potential difference　电势差
resistance　电阻
resistor　电阻器
capacitance　电容
capacitor　电容器
capacitive reactance　容抗
inductance　电感
inductor　电感器
inductive reactance　感抗
impedance　阻抗
sinusoidal voltage　正弦电压
sinusoidal current　正弦电流
wave-length　波长
limiting value　极限值
effective value　有效值
power factor　功率因数
apparent power　视在功率
frequency　频率
initial phase　初相位

phase difference　相位差
efficiency　效率
single-phase　单相
three-phase　三相
neutral conductor　中线
triangular connection　三角形连接
star connection　星形连接
line voltage　线电压
line voltage current　线电流
phase voltage　相电压
phase current　相电流
three-phase three-wire system　三相三线制
rectifier　整流器
full-wave rectifier　全波整流
half-wave rectifier　半波整流
bridge rectifier　桥式整流器
filter　滤波器
pulsating voltage　脉动电压
pulsating current　脉动电流

※ 电磁与电机 ※

magnetic field　磁场
flux　磁通
magnetic flux　磁通量，通量
magnetic field strength, magnetizing force
　　　　　　　磁场强度
magnetic flux density　磁感应强度
reluctance　磁阻
permeability　磁导率
pole　磁极
coil　线圈
winding　绕组
transformer　变压器
core　铁芯
core loss　铁损
eddy current　涡流

turn　线匝
turns ratio　匝数比
coupler　耦合器
self-inductance　自感
mutual inductance　互感
mutual inductor　互感器
induction motor　异步电动机
synchronous motor　同步电动机
generator　发电机
stator　定子
rotor　转子
squirrel-cage rotor　鼠笼式转子
wound rotor　线绕式转子
armature　电枢
self excitation　自励

separately excitation　他励
shunt excitation　并励
series excitation　串励
compounded　复励
load　负载
no-load　空载
full load　满载
starting　启动

starting torque　启动转矩
speed regulation　调速
constant speed　恒速
braking　制动
rated value　额定值
rated power　额定功率
torque　转矩
stepper motor　步进电动机

※ 继电控制与安全用电 ※

switch group　组合开关
button　按钮
relay　继电器
contactor　接触器
thermal over relay　热继电器
time relay　时间继电器
automatic air circuit breaker　自动空气断路器
main circuit　主电路
auxiliary circuit　辅助电路
plug　插头
socket　插座

fuse　熔断器
junction box　接线盒
earthed circuit　接地电路
step voltage　跨步电压
safety voltage　安全电压
electric arc　电弧
glow discharge　辉光放电
explosion-proof　防爆型
flame-proof　隔爆型
intrinsically safe spark　本质安全型火花

※ 电子与传感技术 ※

semiconductor　半导体
transistor　晶体管
silicon control　可控硅，晶闸管
diode　二极管
triode　三极管
light emitting diode(LED)　发光二极管
photodiode　光电二极管
photo-effect transistor(PET)　光敏电阻
thermistor　热敏电阻
emitter　发射极
collector　集电极

base　基极
pulse　脉冲
trigger　触发
cut-off　截止
negative feedback　负反馈
breakdown　击穿
integrated circuit(IC)　集成电路
digital-analog converter　数-模转换器
read only memory(ROM)　只读存储器
random access memory(ROM)　随机存储器
central processing unit(CPU)　中央存储器

sensor　传感器
transducer　传感器；转换器，变频器
piezo-electric transducers　压电式传感器
magnetoelectric transducers　磁电式传感器
thermoelectric transducers　热电式传感器
device　设备，仪器，装置
instrument　仪表
equipment　装备，设备，器械，用具
apparatus　仪器，装置
parameter　参数
data　（datum 的复数）数据

potentiometer　电位计
multimeter　万用表
precision　精度
smartphone　智能手机
drone　雄蜂；无人驾驶飞机
antenna　天线
transmission　传动装置
gigabit　吉（咖）比特
artificial intelligence　人工智能
decision-making system　决策系统

※ 机械零件 ※

part　零件
mechanism　机械，机构
component　元件，构件，部件
executive device　执行机构
slider-crank mechanism　曲柄滑块机构
crank-rocker mechanism　曲柄摇杆机构
design　设计
bolt　螺栓
nut　螺母
screw　螺钉
washer　垫圈
spacer　垫片
thread　线；丝；螺纹
inside thread　内螺纹
belt　皮带
belt pulley　皮带轮
key　键
pin　销
gear　齿轮
gear rack　齿条
spur gear　正齿轮

helical gear　斜齿轮
bevel gear　锥齿轮
spiral bevel gear　螺旋锥齿轮
worm　蜗杆
worm wheel　蜗轮
gearbox　齿轮箱，变速器
shaft　轴
axis　轴，几何轴线，轴心线
bearing　轴承
rolling bearing　滚动轴承
radial bearing　向心轴承
angular contact bearing　角接触轴承
thrust ball bearing　推力球轴承
roller bearing　滚柱轴承
journal bearing　滑动轴承
bushing　衬套；轴衬；轴瓦
coupling　联轴器
clutch　离合器
spring　弹簧，弹性
cam　凸轮，偏心轮
flange　法兰

※ 金属材料与热处理 ※

material 材料
pulling 拉伸
hitting 压缩
shear 剪切
twist 扭转
bending stress 弯曲应力
strength 强度
stress 应力
rigidity,stiffness 刚度
rigidity,hardness 硬度
toughness,ductility 韧性
elastic 弹性的
metal 金属
nonmetal 非金属

steel&iron 钢铁
plain carbon steel 普通碳素钢
cast iron 铸铁
aluminium 铝
alloy 合金
heat treatment 热处理
annealing 退火
spheroidizing 球化退火
normalizing 正火
hardening 淬火
tempering 回火
hardening and tempering 调质
carburizing,cementation 渗碳
carburizing steel,cementation steel 渗碳钢

※ 机械制造工艺 ※

manufacture 制造
manufacturing operations 加工工序
cast 铸造；铸型，模子
mould 模具；铸模；模型
die 冲模，模具
force 凸模，阳模
core 型芯，凹模，阴模
stamping 冲压
punch press 冲床
forge 锻造
weld 焊接，熔接
fusion welding 熔焊
tack welding 点焊
butt welding 对接焊
arc welding 电弧焊

pressure welding 压力焊
gas welding 气焊
weld joint 焊缝
machine tool 机床
machining center 加工中心
NC machine 数控机床
cutting tool,cutter 切削刀具
jig 夹具，钻模；模具
workpiece 工件
table 工作台
turning 车削
lathe 车床
lathe tool 车刀
engine lathe 卧式车床
drilling 钻削

drilling machine, driller　钻床
boring　镗削
reaming　铰孔
milling　铣削
miller, mill　铣床
shaper　铣床,牛头刨床
planning　刨削
grinding　磨削
grinder　磨床
abrasive wheel　砂轮
Machine Coordinate System　机床坐标系
feed　进刀,进给
feed movement　进给运动
cross motion　横向运动
lengthwise movement　纵向运动

tap　丝锤
threading　车螺纹
processing parameters　工艺参数
dimension　尺寸
diameter　直径
nominal diameter　公称直径
metric　公制的;米制的
clearance fit　间隙配合
interference fit　过盈配合
transition fit　过渡配合
hole-based system　基孔制
shaft-based system　基轴制
tolerance　公差,容许偏差
interchangeable　可互换的

※ 液压设备与工程机械 ※

viscometer　黏度计
hydraulic　液压的,液力的
hydraulic cylinder　液压缸
hydraulic valve　液压阀
relief valve　安全阀
hydraulic slave motor　液压马达
hydraulic pump　液压泵
vane wheel　叶轮
vane type pump　叶轮泵

piston　活塞
air compressor　空压机
belt conveyors　皮带运输机,胶带运输机
hoist, haulage　提升机,卷扬机
crane　起重机
internal combustion engine　内燃机
diesel engine　柴油机
speed reducer, reductor　减速器

※ 设备维修 ※

maintain　维护,维修
adjustment　调整,调节;校正

locksmith　钳工
bench　(钳工等的)工作台

assemble　安装
disassemble　拆卸,解体
spanner　扳手
wrench　扭转
tongs　钳子
inspection　检查
lubrication　润滑
oil　润滑油
oil seal　油封

friction　摩擦
wear out　磨损,损耗
rust　腐蚀
vibration　颤动;振动
damping　阻尼
technical requirement　技术要求
slide caliper　游标卡尺
micrometer caliper　千分尺